T0226251

Synthesis Lectures on Computer Science

The series publishes short books on general computer science topics that will appeal to advanced students, researchers, and practitioners in a variety of areas within computer science.

Max Cohen · Calin Belta

Adaptive
and Learning-Based Control
of Safety-Critical Systems

 Springer

Max Cohen
Department of Mechanical Engineering
Boston University
Boston, MA, USA

Calin Belta
Department of Mechanical Engineering
Boston University
Boston, MA, USA

ISSN 1932-1228   ISSN 1932-1686 (electronic)
Synthesis Lectures on Computer Science
ISBN 978-3-031-29312-2   ISBN 978-3-031-29310-8 (eBook)
https://doi.org/10.1007/978-3-031-29310-8

© The Editor(s) (if applicable) and The Author(s), under exclusive license to Springer Nature
Switzerland AG 2023
This work is subject to copyright. All rights are solely and exclusively licensed by the Publisher, whether the whole
or part of the material is concerned, specifically the rights of translation, reprinting, reuse of illustrations, recitation,
broadcasting, reproduction on microfilms or in any other physical way, and transmission or information storage
and retrieval, electronic adaptation, computer software, or by similar or dissimilar methodology now known or
hereafter developed.
The use of general descriptive names, registered names, trademarks, service marks, etc. in this publication does
not imply, even in the absence of a specific statement, that such names are exempt from the relevant protective
laws and regulations and therefore free for general use.
The publisher, the authors, and the editors are safe to assume that the advice and information in this book are
believed to be true and accurate at the date of publication. Neither the publisher nor the authors or the editors give
a warranty, expressed or implied, with respect to the material contained herein or for any errors or omissions that
may have been made. The publisher remains neutral with regard to jurisdictional claims in published maps and
institutional affiliations.

This Springer imprint is published by the registered company Springer Nature Switzerland AG
The registered company address is: Gewerbestrasse 11, 6330 Cham, Switzerland

To my mom

—Max Cohen

To Ana and Stefan

—Calin Belta

Preface

Motivation and Objectives

The rising levels of autonomy exhibited by complex cyber-physical systems have brought questions related to safety and adaptation to the forefront of the minds of controls and robotics engineers. Often, such autonomous systems are deemed to be *safety-critical* in the sense that failures during operation could significantly harm the system itself, other autonomous systems, or, in the worst-case, humans interacting with such a system. Complicating the design of control and decision-making algorithms for safety-critical systems is that they must cope with large amounts of uncertainty as they are deployed autonomously in increasingly real-world environments. For example, a drone required to deliver medicine to remote regions may encounter unknown wind gusts along its path; due to unforeseen weather conditions, an autonomous vehicle may drive on surfaces where the friction forces between the tires and ground become uncertain; a robotic manipulator operating in close proximity with humans may need to transport various loads with uncertain masses while avoiding collisions with humans moving in an unpredictable manner. In all of these scenarios, the autonomous systems' control/decision-making policies must be able to *adapt* to uncertainties while adhering to safety-critical constraints.

Recent advances in artificial intelligence (AI) and machine learning (ML) have facilitated the design of control and decision-making policies directly from data. For example, advancements in reinforcement learning (RL) have enabled robots to learn high-performance control policies purely from trial-and-error interaction with their environment; advances in deep neural network architectures have allowed for learning control policies directly from raw sensory information; advancements in Bayesian inference have allowed for constructing non-parametric models of complicated dynamical systems with probabilistic accuracy guarantees. The fact that these ML techniques can learn control policies and dynamic models directly from data makes them extremely attractive for use in autonomous systems that must operate in the face of uncertainties. Despite these promises, the performance of such ML approaches is tightly coupled to the data used to train the ML model and may act unexpectedly when exposed to data outside its training distribution. That is, although such ML models are extremely expressive at describing the complex input-output relation of the training data, they are not necessarily *adaptive* in that input data outside of the range of the training dataset may produce unexpected outputs. This phenomenon makes it challenging to directly deploy such learning-based controllers on safety-critical systems that will inevitably encounter unexpected scenarios that cannot be accounted for using pre-existing data.

The main objective of this book is to present a unified framework for the design of controllers that learn from data online with formal guarantees of correctness. We are primarily concerned with ensuring that such learning-based controllers provide *safety* guarantees, a property formalized using the framework of set invariance. Our focus in this book is on *online* learning-based control, or *adaptive control*, in which learning and control occur simultaneously in the feedback loop. Rather than using a controller trained on an a priori dataset collected offline that is then statically deployed on a system, we are interested in using real-time data to continuously update the control policy online and cope with uncertainties that are challenging to characterize until deployment. In this regard, most of the controllers developed in this book are *dynamic* feedback controllers in that they depend on the states of an auxiliary dynamical system representing an adaptation algorithm that evolves based upon data observed in real-time. This idea is not new—it has been the cornerstone of the field of adaptive control for decades. From this perspective, the main objective of this book is to extend techniques from the field of adaptive control, which has primarily been concerned with stabilization and tracking problems, to consider more complex control specifications, such as safety, that are becoming increasingly relevant in the realm of robotic and autonomous systems.

Intended Audience

This book is intended to provide an introduction to learning-based control of safety-critical systems for a wide range of scientists, engineers, and researchers. We have attempted to write this book in a self-contained manner—a solid background in vector calculus, linear algebra, and differential equations should be sufficient to grasp most of the mathematical concepts introduced herein. Prior knowledge of nonlinear systems theory would be useful (e.g., an introductory course that covers the basics of Lyapunov stability) as it serves as the starting point for most of the developments in this book, but is not strictly necessary as we briefly cover the concepts used throughout this book in Chap. 2. Researchers from control theory are shown how established control-theoretic tools, such as Lyapunov functions, can be suitably transposed to address problems richer than stabilization and tracking problems. They are also exposed to ML and its integration with control-theoretic tools with the goal of dealing with uncertainty. ML researchers are shown how control-theoretic and formal methods tools can be leveraged to provide guarantees of correctness for learning-based control approaches, such as reinforcement learning.

Book Outline and Usage

This book is organized into nine chapters. In most chapters, we begin with a short introduction that provides motivation and an overview of the methods introduced therein and then immediately move into the technical content. Discussions on related works are postponed until the *Notes* section, which can be found at the end of each chapter. We have aimed to prove most of the results we state; however, in an effort to keep this book relatively self contained, we omit proofs that require the reader to consult outside sources, and instead provide reference to where such proofs may be found in the *Notes* section of each chapter. For increased readability, we do not cite references in the technical parts—the works on which the material is based are cited in the *Notes* sections at the end of each chapter. The contents of each chapter are summarized as follows:

- In Chap. 1 we provide an informal overview of the topics discussed in this book.
- In Chap. 2 we review the fundamentals of Lyapunov stability theory and how such ideas can be directly used for control synthesis using the notion of a control Lyapunov function (CLF). Although these concepts are likely familiar to many readers, here we recast these ideas in the modern context of optimization-based control in which control inputs are computed as the solution to convex optimization problems that guarantee stability of the closed-loop system by construction.
- In Chap. 3 we provide a concise introduction to the field of safety-critical control in which the central object of study is a *control barrier function* (CBF), an extension of CLFs from stability to safety problems. Before introducing CBFs, we first review the idea of formalizing the concept of safety in dynamical systems using the notion of set invariance. After covering CBFs, we present an extension of the CBF methodology using the notion of a *high order CBF* (HOCBF), which provides a systematic framework for dynamically extending a CBF candidate to construct a valid safety certificate.
- In Chap. 4, we provide a short introduction to adaptive control of nonlinear systems—a field focused on simultaneous learning and control of uncertain nonlinear systems. Our discussion on nonlinear adaptive control centers around the notion of an *adaptive CLF* (aCLF), which extends the CLF paradigm from Chap. 2 to adaptive control systems. Here, we also introduce the more modern adaptive control concept of *concurrent learning*—a data-driven technique that can be used to strengthen traditional adaptive controllers.
- In Chap. 5 we unite the safety-critical control framework of Chap. 3 with the nonlinear adaptive control framework proposed in Chap. 4 via the notion of an *adaptive CBF* (aCBF). In particular, we demonstrate how the concurrent learning adaptive control techniques from the previous chapter can be used to enforce safety under the worst-case bounds on the model uncertainty while simultaneously reducing such bounds to allow for less conservative behavior as more data about the system is collected.

- In Chap. 6 we extend the adaptive control methods from Chaps. 4 to 5 to a larger set of parameter estimation algorithms using the notions of input-to-state stability and input-to-state-safety. The adaptive control techniques in this chapter are referred to as "modular" as they allow interchangeability of the parameter estimation algorithm without affecting the stability and safety guarantees of the adaptive control.

- In Chap. 7 we extend the adaptive control techniques from previous chapters to handle a more general class of uncertainties. In particular, whereas the methods from Chaps. 4 to 6 handled systems in which the uncertainty enters the dynamics in an additive fashion, in this chapter we consider systems in which the uncertainty enters the dynamics in a multiplicative fashion in the sense that the dynamics are bilinear in the control and uncertainty.

- In Chap. 8, we move from an adaptive control framework to a reinforcement learning (RL) framework in which the goal is to control a system to optimize a cost function while satisfying safety-critical constraints. We illustrate how the adaptive control techniques from earlier chapters can be extended to this domain by developing safe *online* model-based RL (MBRL) algorithms that allow for safely learning the system dynamics and solution to an optimal control problem online in real-time, rather than in an episodic learning framework as is typical in RL approaches.

- In Chap. 9, we broaden the class of control specifications under consideration. In particular, we shift our objective from designing controllers that guarantee stability and safety to controllers that guarantee the satisfaction of more general *linear temporal logic* (LTL) specifications. Here, we show how the problem of LTL controller synthesis can be broken down into a sequence of reach-avoid problems that can be solved using the Lyapunov and barrier functions introduced in earlier chapters. As a specific example of this idea, we apply the MBRL framework from Chap. 8 to solve such a sequence of reach-avoid problems, and ultimately, to design a controller that enforces satisfaction of an LTL specification.

This book can be used and read in a few different ways. Although it was written primarily to present research results by the authors and others in a unified framework, it could also serve as the main reference for teaching a course for various audiences. For an audience with a strong background in control theory, the material in this book could be used to teach a short course that provides an overview of safe learning-based control from an adaptive control perspective. For an audience with less exposure to advanced topics in control theory (e.g., first-year graduate students), this book could be used to teach a full course on the topic of safe learning-based control by covering the material in this book along with more extensive coverage of background material in nonlinear systems, adaptive control, machine learning, and formal methods.

Boston, MA, USA   Max Cohen
                                                                                                          Calin Belta

Acknowledgements We gratefully acknowledge our collaborators who contributed to the results in this book: we thank Roberto Tron who was instrumental in developing the approach detailed in Chap. 7, and we thank Zachary Serlin and Kevin Leahy who contributed to the results reported in Chap. 9. The first author gratefully acknowledges support from the National Science Foundation Graduate Research Fellowship Program, which has funded much of the research presented in this book. The second author acknowledges partial support from the National Science Foundation.

Contents

Acronyms

aCBF	Adaptive control barrier function
aCLF	Adaptive control Lyapunov function
BE	Bellman error
CBF	Control barrier function
CLF	Control Lyapunov function
DBA	Deterministic Büchi automaton
DNN	Deep neural network
DTA	Distance to acceptance
eISS	Exponentially input-to-state stable
eISS-CLF	Exponential input-to-state stabilizing control Lyapunov function
ES-aCLF	Exponentially stabilizing adaptive control Lyapunov function
ES-CLF	Exponentially stabilizing control Lyapunov function
FE	Finite excitation
HJB	Hamilton-Jacobi-Bellman equation
HOCBF	High order control barrier function
HO-RaCBF	High order robust adaptive control barrier function
ISS	Input-to-state stability
ISSf	Input-to-state safety
ISSf-CBF	Input-to-state safe control barrier function
ISSf-HOCBF	Input-to-state safe high order control barrier function
KKT	Karush-Kuhn-Tucker
LP	Linear program
LQR	Linear quadratic regulator
LTL	Linear temporal logic
MBRL	Model-based reinforcement learning
PE	Persistence of excitation
QP	Quadratic program
RaCBF	Robust adaptive control barrier function
RCBF	Robust control barrier function
RCLF	Robust control Lyapunov function

RL	Reinforcement learning
RLS	Recursive least squares
ROI	Region of interest
SMID	Set membership identification

Notation

$\mathbb{N}$	Set of natural numbers
$\mathbb{Z}$	Set of integers
$\mathbb{R}$	Set of real numbers
$\mathbb{R}_{\geq a}$	Set of real numbers greater than or equal to $a \in \mathbb{R}$
$\mathbb{R}_{>a}$	Set of real numbers strictly greater than $a \in \mathbb{R}$
$\mathbb{R}^n$	The n-dimensional Euclidean vector space
x_i	The ith component of a vector $x \in \mathbb{R}^n$
$x^\top$	Transpose of a vector $x \in \mathbb{R}^n$
$x^\top y$	Inner product $\sum_{i=1}^n x_i y_i$ between two vectors $x, y \in \mathbb{R}^n$
$\|x\|$	Euclidean norm of a vector $x \in \mathbb{R}^n$
$\mathbb{R}^{m \times n}$	Set of $m \times n$ matrices with real entries
A_{ij}	The (i, j) entry of a matrix $A \in \mathbb{R}^{m \times n}$
$A^\top$	Transpose of a matrix $A \in \mathbb{R}^{m \times n}$
$\|A\|$	Induced norm of a matrix $A \in \mathbb{R}^{m \times n}$
$I_{n \times n}$	The $n \times n$ identity matrix
$\lambda_{\min}(A)$	Minimum eigenvalue of a matrix $A \in \mathbb{R}^{n \times n}$
$\lambda_{\max}(A)$	Maximum eigenvalue of a matrix $A \in \mathbb{R}^{n \times n}$
$\emptyset$	Empty set
2^C	Power set of set C
$C_1 \times C_2$	Cartesian product of sets C_1 and C_2
∂C	Boundary of a set C
$\text{Int}(C)$	Interior of a set C
$\bar{C}$	Closure of a set C
$C_1 \setminus C_2$	Set difference of sets C_1 and C_2
$\mathcal{B}_r(x)$	Open ball of radius $r \in \mathbb{R}_{>0}$ centered at $x \in \mathbb{R}^n$
$\bar{\mathcal{B}}_r(x)$	Closed ball of radius $r \in \mathbb{R}_{>0}$ centered at $x \in \mathbb{R}^n$
$\frac{\partial h}{\partial x}(x)$	The $m \times n$ Jacobian matrix of a continuously differentiable function $h : \mathbb{R}^n \to \mathbb{R}^m$ evaluated at $x \in \mathbb{R}^n$
$\nabla h(x)$	The gradient of a continuously differentiable scalar function $h : \mathbb{R}^n \to \mathbb{R}$ evaluated at $x \in \mathbb{R}^n$

$L_f h$	The Lie derivative $L_f h(x) = \frac{\partial h}{\partial x}(x)f(x)$ of a continuously differentiable function $h : \mathbb{R}^n \to \mathbb{R}$ along a vector field $f : \mathbb{R}^n \to \mathbb{R}^n$
$L_g h$	The Lie derivative $L_g h(x) = \frac{\partial h}{\partial x}(x)g(x)$ of a continuously differentiable function $h : \mathbb{R}^n \to \mathbb{R}$ along a vector field $g : \mathbb{R}^n \to \mathbb{R}^{n \times m}$
$K : \mathcal{X} \rightrightarrows \mathcal{U}$	A set-valued mapping that assigns to each $x \in \mathcal{X}$ a set $K(x) \subset \mathcal{U}$
$\mathcal{K}$	Set of class $\mathcal{K}$ functions
$\mathcal{K}_\infty$	Set of class $\mathcal{K}_\infty$ functions
$\mathcal{KL}$	Set of class $\mathcal{KL}$ functions
$\mathcal{K}^e$	Set of extended class $\mathcal{K}$ functions
$\mathcal{K}_\infty^e$	Set of extended class $\mathcal{K}_\infty$ functions
$\mathcal{L}_\infty$	Space of piecewise continuous and bounded functions
$\mathcal{L}_2$	Space of piecewise continuous and square integrable functions
$w_o(0)w_o(1)\dots w_o(i) \in O$	Word over set O
$(w_o(0)w_o(1)\dots w_o(k))^\omega$	Infinitely many repetitions of a sequence
O^ω	Set of infinite words over set O
$\top$	Boolean constant true
$\neg, \wedge, \vee, \to, \leftrightarrow$	Boolean operators negation, conjunction, disjunction, implication, and equivalence
$\bigcirc$	"Next" temporal operator
U	"Until" temporal operator
$\Diamond$	"Eventually" temporal operator
$\Box$	"Always" temporal operator

Introduction

In this brief introductory chapter, we provide an informal description of the problem that we consider throughout the book. We introduce the two classes of control systems that we focus on and discuss our assumptions and relevance to applications. We also provide a high-level description of the technical approach.

Safe learning-based control generally concerns the design of control policies for uncertain dynamical systems that ensure the satisfaction of safety constraints. Such safety constraints may be represented by constraints on the states of a dynamical system (i.e., the system trajectory should remain within a prescribed "safe set" at all times), constraints on the actuators of a system (i.e., control effort must not exceed specified bounds), or both. Secondary to such safety requirements are performance requirements, such as asymptotic stability of an equilibrium point or the minimization of a cost function, that would be desirable to satisfy, but not at the cost of violating any safety constraints. The challenge that safe learning-based control aims to address is the construction of control policies enforcing the satisfaction of these aforementioned requirements in the face of model uncertainty. To make these ideas more concrete, consider the control system

$$\dot{x} = f(x, u),$$

where $x \in X$ is the system state belonging to the state space X, $u \in \mathcal{U}$ is the control input belonging to the (possibly bounded) control space $\mathcal{U}$, and $f(x, u)$ is a vector field characterizing the system dynamics. The main problem considered herein is to design a control policy that renders the above system *safe*, that is, that the resulting closed-loop trajectory $x(t)$ remains in a safe set $x(t) \in C \subset X$ at all times under the control input $u(t) \in \mathcal{U}$, without (precise) knowledge of the vector field f governing the system dynamics. In this book, we present a suite of tools for solving this problem by building an estimate of the dynamics $\hat{f}(x, u)$ using data collected *online* that is then used in a model-based control policy

© The Author(s), under exclusive license to Springer Nature Switzerland AG 2023
M. Cohen and C. Belta, *Adaptive and Learning-Based Control of Safety-Critical Systems*,
Synthesis Lectures on Computer Science,
https://doi.org/10.1007/978-3-031-29310-8_1

to satisfy the aforementioned objectives. Importantly, we aim to ensure that such a policy guarantees satisfaction of all safety constraints not only after a suitable estimate has been obtained, but also *during* the learning process. Such a problem is very challenging without further restrictions on the system model or learning approach used—in what follows we discuss a special case of this more general problem and outline the main techniques used to solve it.

The primary focus of this book is on safe learning-based control from an adaptive control perspective. For much of this book, our development focuses on nonlinear control affine systems with parametric uncertainty:

$$\dot{x} = f(x) + F(x)\theta + g(x)u,$$

where $\theta \in \Theta$ is a vector of uncertain parameters belonging to the (possibly bounded) parameter space Θ. Here, $f(x)$ models the system drift dynamics describing the natural flow of the system in the absence of control inputs or uncertainty, $g(x)$ is a matrix whose columns capture the system's control directions, and $F(x)$ is a matrix whose columns characterize the directions along which the uncertain parameters act. The assumption that the uncertainty manifests itself in a parametric fashion implies that the *structure* of the dynamics are known but may depend on various quantities, such as inertial or friction parameters, that are unknown. Fortunately, most real-world robotic systems (e.g., those whose equations of motion are derived using Lagrangian mechanics) satisfy this structural assumption, and systems not satisfying such an assumption can often be made to satisfy it using hand-picked or learned features to represent the uncertainty.

We are primarily interested in designing adaptive controllers $u = k(x, \hat{\theta})$, where $\hat{\theta}$ is an estimate of the uncertain parameters, that ensure the closed-loop system trajectory $x(t)$ remains in some prescribed safe set $C \subset X$ at all times. Complicating this problem is the fact that $\hat{\theta}$ is not a static estimate of the uncertainty, but continuously evolves according to its own dynamics $\dot{\hat{\theta}}$, which may, in turn, depend on the system state x and even the control input u. This gives rise to a tight feedback loop between learning and control, which necessitates carefully selecting both the control policy $k(x, \hat{\theta})$ and the learning algorithm itself—characterized by the auxiliary dynamical system $\dot{\hat{\theta}}$—to ensure the ultimate control objective is met. To this end, we demonstrate how modern ideas from machine learning can be incorporated into traditional adaptation algorithms to guarantee convergence of the parameter estimates to their true values, allowing for a gradual reduction in uncertainty as more data about the system is collected.

We approach the problem of designing learning-based control algorithms for safety-critical systems through the use of *certificate functions* from nonlinear control theory. The most familiar of such functions is the well-known *Lyapunov function* for certifying the stability of dynamical systems, and its extension to control systems—the *control Lyapunov function* (CLF). The main ideas regarding Lyapunov can be suitably transposed to address safety, rather than stability, using the dual notion of a *barrier function* for safety certification, and a *control barrier function* (CBF) for control synthesis. Much of this book concentrates on developing adaptive versions of CLFs and CBFs that facilitate the design of both controllers and adaptation algorithms that guarantee stability and safety, respectively, by construction. Here, we recast the traditional design of adaptive controllers in the modern context of optimization-based control in which control inputs are computed as the solution to a convex optimization problem whose constraints guarantee satisfaction of closed-loop system properties (stability, safety) by construction.

In later parts of this book, we shift from the problem of safe adaptive control to *safe reinforcement learning* in which the objective is to construct a controller for an uncertain dynamical system that minimizes the infinite-horizon cost functional

$$\int_0^\infty \ell(x(s), u(s))ds,$$

where $\ell(x, u)$ is a running cost, while ensuring that the system state remains in a safe set at all times. Extending the adaptive control ideas introduced in early chapters, we demonstrate how similar techniques can be leveraged to safely learn a parametric estimate of the value function of the above optimal control problem online using data from a single trajectory.

Finally, we discuss an extension of the aforementioned approaches to richer control specifications given in the form of *temporal logic* formulas, which provide a formal way to express complex control objectives beyond that of stability and safety. We focus on Linear Temporal Logic (LTL), which is particularly fitted for synthesis of control strategies using automata-based techniques. We show how the safety guarantees developed for adaptive control throughout the book can be extended to accommodate expressive temporal logic specifications.

Stabilizing Control Design

2

In the first technical chapter of the book, we introduce fundamental notions of stability for dynamical systems, and review ways in which stability can be enforced through feedback control. Central to our treatment is the notion of a control Lyapunov function, which maps stability requirements to control constraints. The most important statement that we make in this chapter is that stability can be enforced as a constraint in an optimization problem, which paves the way to seamless integration with safety, which is treated later in the book. This chapter is organized as follows. In Sect. 2.1, we define stability and Lyapunov functions, and review the main stability verification results based on Lyapunov theory. We introduce control Lyapunov functions in Sect. 2.2, where we also discuss enforcing stability as a constraint in an optimization problem. We present two widely used methods for designing control Lyapunov functions in Sect. 2.3. We conclude with references, discussions, and suggestions for further reading in Sect. 2.4.

2.1 Lyapunov Stability Theory

Here we recount the main ideas behind Lyapunov stability theory, which serves as a foundation for the technical developments in later chapters. Much of this material may be familiar to many readers; however, we will use it as a means to establish common ground regarding notation, definitions, and preliminary results that will be used throughout the rest of the book. Moreover, we hope this presentation will help highlight the duality between stability and safety properties, encoded via Lyapunov functions and barrier functions, respectively, the latter of which will be covered in the proceeding chapters.

© The Author(s), under exclusive license to Springer Nature Switzerland AG 2023
M. Cohen and C. Belta, *Adaptive and Learning-Based Control of Safety-Critical Systems*,
Synthesis Lectures on Computer Science,
https://doi.org/10.1007/978-3-031-29310-8_2

2.1.1   Stability Notions

We consider a nonlinear dynamical system described by

$$\dot{x} = f(x), \tag{2.1}$$

where $x \in X \subset \mathbb{R}^n$ is the system state, assumed to take values in an open subset $X \subset \mathbb{R}^n$ of the n–dimensional Euclidean space, and $f : X \to \mathbb{R}^n$ is a vector field that is locally Lipschitz on X, which models the system dynamics.

Definition 2.1   A mapping $f : X \to Y$ with $X \subset \mathbb{R}^n$ and $Y \subset \mathbb{R}^m$ is said to be locally Lipschitz on X if, for all $x_0 \in X$, there exist positive constants $\delta, L \in \mathbb{R}_{>0}$ such that for all $x, y \in \mathcal{B}_\delta(x_0)$,

$$\|f(x) - f(y)\| \leq L\|x - y\|. \tag{2.2}$$

When $f : X \to Y$ is locally Lipschitz on X and the domain of the mapping X is understood from the context, we will simply say that f is locally Lipschitz.

The Lipschitz assumption on a vector field f ensures that the associated dynamical system generates unique trajectories from any initial condition $x_0 \in X$, at least for short time intervals. Formally, we have:

Theorem 2.1   *Let $f : X \to \mathbb{R}^n$ be locally Lipschitz. Then, for any initial condition $x_0 \in X$, there exists a maximal interval of existence $I(x_0) := [0, \tau_{\max}) \subset \mathbb{R}$, $\tau_{\max} \in \mathbb{R}_{>0}$, and a continuously differentiable mapping $x : I(x_0) \to X$ such that $t \mapsto x(t)$ is the unique solution to (2.1) on $I(x_0)$ in the sense that it solves the initial value problem*

$$\begin{aligned} x(0) &= x_0, \\ \dot{x}(t) &= f(x(t)), \quad \forall t \in I(x_0). \end{aligned} \tag{2.3}$$

The vector field f is called *forward complete* if $\tau_{\max} = \infty$ in the above theorem. With a slight abuse of notation, we use $x(\cdot)$ or $t \mapsto x(t)$ to distinguish a trajectory of (2.1) from an arbitrary state of (2.1), which is simply denoted by x.

A fundamental problem studied in control theory concerns the stability of equilibrium points of (2.1).

Definition 2.2   A point $x_e \in X$ is said to be an equilibrium point of (2.1) if $f(x_e) = 0$.

For the remainder of this chapter, we assume that there exists at least one equilibrium point of (2.1), which, without loss of generality, is assumed to be at the origin. Informally, the origin is said to be stable for (2.1) if initial conditions near the origin produce trajectories that stay near it. The origin is asymptotically stable if such trajectories also approach the origin. One way to formally define various notions of stability is through the use of different classes of scalar *comparison functions*. Two classes of comparison functions useful for studying stability are defined as follows.

Definition 2.3 (*Class $\mathcal{K}$, $\mathcal{K}_\infty$ functions*) A continuous function $\alpha : [0, a) \to \mathbb{R}_{\geq 0}$, where $a \in \mathbb{R}_{>0}$, is said to be a *class $\mathcal{K}$ function*, denoted by $\alpha \in \mathcal{K}$, if $\alpha(0) = 0$ and $\alpha(\cdot)$ is strictly increasing. If $a = \infty$ and $\lim_{r \to \infty} \alpha(r) = \infty$, then α is said to be a *class $\mathcal{K}_\infty$ function*, which we denote by $\alpha \in \mathcal{K}_\infty$.

Definition 2.4 (*Class $\mathcal{KL}$ function*) A continuous function $\beta : [0, a) \times \mathbb{R}_{\geq 0} \to \mathbb{R}_{\geq 0}$, where $a \in \mathbb{R}_{>0}$, is said to be a class $\mathcal{KL}$ function, denoted by $\beta \in \mathcal{KL}$, if $\beta(r, s)$ is a class $\mathcal{K}$ function of r for each fixed $s \in \mathbb{R}_{\geq 0}$, and if $\beta(r, s)$ is decreasing in s for each fixed $r \in [0, a)$.

Making use of these comparison functions allows us to concisely define various notions of stability.

Definition 2.5 (*Stability*) Let $t \mapsto x(t)$ be a trajectory of (2.1) from an initial condition $x_0 \in X$ in the sense of (2.3). The origin for (2.1) is said to be

- *locally stable* if there exists $\alpha \in \mathcal{K}$ and a positive constant $\delta \in \mathbb{R}_{>0}$, such that for all $x_0 \in \mathcal{B}_\delta(0)$,

$$\|x(t)\| \leq \alpha(\|x_0\|), \quad \forall t \in \mathbb{R}_{\geq 0}; \tag{2.4}$$

- *locally asymptotically stable* if there exists $\beta \in \mathcal{KL}$ and a positive constant $\delta \in \mathbb{R}_{>0}$, such that for all $x_0 \in \mathcal{B}_\delta(0)$,

$$\|x(t)\| \leq \beta(\|x_0\|, t), \quad \forall t \in \mathbb{R}_{\geq 0}; \tag{2.5}$$

- *locally exponentially stable* if it is locally asymptotically stable and $\beta(r, s) = kre^{-cs}$ for some positive constants $k, c \in \mathbb{R}_{>0}$.

If the above conditions hold for all $x_0 \in X$ then we remove the "local" qualifier and simply say that the origin is stable, asymptotically stable, or exponentially stable.

According to the above definitions, the stability of an equilibrium point implies that the trajectory from a fixed initial condition x_0 will remain within a ball of radius $\alpha(\|x_0\|)$ centered at the origin. If the system is asymptotically stable, then, in addition to remaining in some ball of the origin, the properties of class $\mathcal{KL}$ functions allow us to conclude that trajectories approach the origin in the limit as time goes to infinity. Thus, stability can be used to answer questions related to the safety ("never do anything bad") and liveness ("eventually do something good") properties of a system. For example, if the ball of radius $\alpha(\|x_0\|)$ centered at the origin represents some "safe" region that the system should remain in at all times, then certifying the stability of (2.1) allows one to conclude safety of the system[1]. On the other hand, if the origin represents some desired state that the system should reach, then certifying asymptotic stability allows one to conclude that the system will eventually converge to some neighborhood of that state. If asymptotic stability can be strengthened to exponential, then one can even provide estimates on how long it takes for the system to reach such a region. The main limitation of the preceding approach is that certifying these properties depends on knowledge of the system trajectories, which is problematic since nonlinear dynamical systems of the form (2.1) generally do not admit closed-form solutions.

2.1.2 Lyapunov Functions

One way to certify the stability of the equilibrium points of a dynamical system without explicit knowledge of its trajectories is through the use of a Lyapunov function. Roughly speaking, a Lyapunov function is a positive definite scalar function that captures a generalized notion of the system's "energy"; if this energy is conserved or decays along system trajectories, then one can draw conclusions about stability. To examine how a scalar function changes along a system's trajectories without explicitly requiring knowledge of such trajectories, we use the *Lie derivative*. Given a scalar function $V : X \to \mathbb{R}$ and a vector field $f : X \to \mathbb{R}^n$, the Lie derivative of V along f is defined as

$$L_f V(x) := \frac{\partial V}{\partial x}(x) f(x) = \nabla V(x)^\top f(x), \tag{2.6}$$

which measures the rate of change of V along the vector field f. The Lie derivative directly encodes the time rate of change of V along a given trajectory $t \mapsto x(t)$ since

$$\frac{d}{dt} V(x(t)) = \frac{\partial V}{\partial x}(x(t))\dot{x}(t) = \frac{\partial V}{\partial x}(x(t)) f(x(t)) = L_f V(x(t)).$$

The notion of a Lie derivative generalizes to vector-valued functions $h : \mathbb{R}^n \to \mathbb{R}^m$ using the same definition as in (2.6), in which case $L_f h(x) = \frac{\partial h}{\partial x}(x) f(x) \in \mathbb{R}^m$ will be a m-dimensional vector.

Since $\dot{V}$ denotes differentiation of V with respect to time, we often abbreviate the Lie derivative as $\dot{V}(x) := L_f V(x)$. The main idea behind certifying stability using the Lyapunov method is to guarantee that some positive definite scalar "energy" function V decreases along system trajectories, which can be ensured by checking that

$$L_f V(x) < 0 \tag{2.7}$$

---

[1] A more in-depth treatment of safety is postponed to the next chapter.

for all $x \in X \setminus \{0\}$. If such a condition holds and V is positive definite, then $V(x(t))$ must decrease along any solution $t \mapsto x(t)$ to zero, implying the solution itself converges to zero. If such an energy function V satisfies the preceding conditions (positive definiteness, negative definite Lie derivative), then we refer to V as a *Lyapunov function*, which provides a certificate for the stability of the system. To formalize these ideas, we give a few preliminary definitions.

Definition 2.6 (*Lyapunov function candidate*) A continuously differentiable scalar function $V : X \to \mathbb{R}_{\geq 0}$ is said to be a *Lyapunov function candidate* if there exist $\alpha_1, \alpha_2 \in \mathcal{K}_\infty$, such that for all $x \in X$,

$$\alpha_1(\|x\|) \leq V(x) \leq \alpha_2(\|x\|). \tag{2.8}$$

The requirement that a Lyapunov function candidate be bounded by class $\mathcal{K}_\infty$ functions can be relaxed to require that $\alpha_1, \alpha_2 \in \mathcal{K}$ if one only wishes to only establish local stability results. The existence of such class $\mathcal{K}$ functions for all $x \in X$ is guaranteed if $V : X \to \mathbb{R}_{\geq 0}$ is positive definite[2] on X. If, in addition, V is radially unbounded[3], then α_1, α_2 can be taken as class $\mathcal{K}_\infty$ functions. For ease of exposition, our working definition of a Lyapunov function candidate will assume that $\alpha_1, \alpha_2 \in \mathcal{K}_\infty$ with the understanding that all stability results can be given a local characterization through the use of class $\mathcal{K}$ functions.

Definition 2.7 (*Lyapunov function*) A Lyapunov function candidate is said to be a *Lyapunov function* if

$$\dot{V}(x) = L_f V(x) < 0, \quad \forall x \in X \setminus \{0\}. \tag{2.9}$$

Note that if V is a Lyapunov function, then $L_f V$ is negative definite on X, implying the existence of $\alpha \in \mathcal{K}$ such that

$$L_f V(x) \leq -\alpha(\|x\|), \quad \forall x \in X.$$

We now have all the tools in place to state the main result regarding Lyapunov functions.

Theorem 2.2 (Lyapunov's Direct Method) *Let the origin be an equilibrium point of* (2.1) *and let* $V : X \to \mathbb{R}_{\geq 0}$ *be a Lyapunov function. Then, the origin of* (2.1) *is asymptotically stable.*

The power of Theorem 2.2 is that it allows for certifying stability purely based on the vector field f describing the system dynamics (2.1) and a suitably constructed Lyapunov function V: verifying that f points directly into the sublevel sets of the Lyapunov function V

---

[2] Recall that a function $V : X \to \mathbb{R}_{\geq 0}$ is positive definite on X if $V(x) > 0$ for all $x \in X \setminus \{0\}$ and $V(x) = 0$ if and only if $x = 0$.
[3] Recall that a function is radially unbounded if $\lim_{\|x\| \to \infty} V(x) = \infty$.

is sufficient to certify the stability of the equilibrium point. If stronger conditions are placed on V, then such an approach can be used to certify the exponential stability of equilibria.

Theorem 2.3 *Let the conditions of Theorem 2.2 hold and suppose that $\alpha_i(\|x\|) = c_i\|x\|^2$ with $c_i \in \mathbb{R}_{>0}$ for all $i \in \{1, 2, 3\}$. Then, the origin of (2.1) is exponentially stable.*

We will use the preceding theorem as an opportunity to introduce a tool known as the *Comparison Lemma* that allows for bounding the trajectories of a possibly very complicated dynamical system by the trajectories of a much simpler one.

Lemma 2.1 (Comparison Lemma) *Let $f : \mathbb{R} \times \mathbb{R}_{\geq 0} \to \mathbb{R}$ be locally Lipschitz in its first argument and continuous in its second. Consider the initial value problem*

$$\dot{y}(t) = f(y(t), t), \quad \forall t \in I(y_0),$$
$$y(t_0) = y_0,$$

where $I(y_0) = [t_0, \tau_{\max})$ is the solution's maximal interval of existence. Now let $v : [t_0, \tau_{\max}) \to \mathbb{R}$ be a continuously differentiable function satisfying

$$\dot{v}(t) \leq f(v(t), t), \quad \forall t \in [t_0, \tau_{\max})$$
$$v(t_0) \leq y_0.$$

Then $v(t) \leq y(t)$ for all $t \in [t_0, \tau_{\max})$.

We now demonstrate how the Comparison Lemma in conjunction with Theorem 2.3 can be used to establish exponentially decaying bounds on the trajectory of a dynamical system.

Proof (of Theorem 2.3) The Lie derivative of the Lyapunov function candidate can be bounded as

$$\dot{V}(x) = L_f V(x) \leq -c_3\|x\|^2 \leq -\frac{c_3}{c_2}V(x).$$

We now introduce the scalar comparison system

$$\dot{y} = -\frac{c_3}{c_2}y$$
$$y(0) = V(x_0),$$

whose solution is given by

$$y(t) = V(x_0)e^{-\frac{c_3}{c_2}t}.$$

It then follows from the Comparison Lemma that

$$V(x(t)) \leq V(x_0)e^{-\frac{c_3}{c_2}t}.$$

Using $c_1 \|x\|^2 \leq V(x) \leq c_2 \|x\|^2$, the above inequality implies that

$$\|x(t)\| \leq \sqrt{\frac{c_2}{c_1}} \|x_0\| e^{-\frac{c_3}{2c_2} t},$$

which implies the origin is exponentially stable, as desired.   □

The main limitation of the Lyapunov approach is that it relies on constructing a Lyapunov function, which often raises the question "How does one find a Lyapunov function?" If no additional assumptions are placed on f (other than those required for the existence and uniqueness of solutions to the corresponding differential equation), then this is a challenging question to answer, and we do not attempt to do so in this book. Rather, we note that in this book the question of "finding a Lyapunov function" is not necessarily the question we seek to answer, since it is implicit in this question that a controller has already been designed for a system, and that our goal is to verify that such a controller renders the origin stable for the closed-loop system. If the ultimate objective of the control design, however, is to enforce stability, then a different approach would be to design a controller that enforces the Lyapunov conditions by construction, thereby obviating the need to perform post hoc verification or find a Lyapunov function for the closed-loop system. In the following section, we explore this approach using the notion of a *control Lyapunov function*.

2.2   Control Lyapunov Functions

In the previous section, we briefly recounted the main ideas behind Lyapunov stability theory and illustrated how Lyapunov functions can be used to certify the stability of closed-loop dynamical systems of the form $\dot{x} = f(x)$. In this section, we discuss an extension of Lyapunov methods to open dynamical systems, or *control systems*, of the form $\dot{x} = f(x, u)$, where u is a control input that allows for modifying the natural dynamics of the system, encoded by the vector field $f(x, 0)$. Note that by fixing a feedback controller $u = k(x)$, we obtain a new closed-loop dynamical system $\dot{x} = f_{cl}(x) := f(x, k(x))$ whose stability properties could then be studied by searching for a suitable Lyapunov function as in Sect. 2.1. In this section, we present an alternative approach to certifying the stability of control systems based on the notion of a *control Lyapunov function* (CLF). In essence, a CLF is a Lyapunov function candidate for a control system, whose Lie derivative can be made to satisfy the Lyapunov conditions by appropriate control action.

The main idea behind the CLF approach is as follows. Rather than fixing a desired controller and then searching for a Lyapunov function, we fix a desired Lyapunov function and then construct a controller such that the Lyapunov conditions are satisfied for the closed-loop system by construction. The distinction between these two approaches is subtle, but we believe important. The process of finding a Lyapunov function is very challenging, often requiring a certain amount of ingenuity or trial and error. However, the process of finding a

CLF can be made almost systematic for many relevant classes of control systems. Various methods to construct CLFs are reviewed in Sect. 2.3.

We focus on a class of nonlinear systems of the form

$$\dot{x} = f(x) + g(x)u, \tag{2.10}$$

where $x \in X \subset \mathbb{R}^n$ is the state of the system with X an open set, and $u \in \mathcal{U} \subset \mathbb{R}^m$ is the control input. The vector field $f : X \to \mathbb{R}^n$ models the system *drift dynamics* and is assumed to be locally Lipschitz, whereas the columns of $g : X \to \mathbb{R}^{n \times m}$, denoted by $g_i : X \to \mathbb{R}^n$, $i \in \{1, \dots, m\}$, are vector fields capturing the system's *control directions*, also assumed to be locally Lipschitz. Control systems in the form (2.10) are called *affine* since the right-hand side of (2.10) is an affine function in control (if the state is fixed).

Given a state feedback controller $u = k(x)$, we obtain the corresponding closed-loop system:

$$\dot{x} = f(x) + g(x)k(x) =: f_{\mathrm{cl}}(x). \tag{2.11}$$

Note that if $k : X \to \mathcal{U}$ is locally Lipschitz, then the corresponding closed-loop vector field $f_{\mathrm{cl}} : X \to \mathbb{R}^n$ is also locally Lipschitz. Hence, by Theorem 2.1, for any initial condition $x_0 \in X$, there exists a maximal interval of existence $I(x_0) = [0, \tau_{\max}) \subseteq \mathbb{R}_{\geq 0}$ and a continuously differentiable function $x : I(x_0) \to X$ such that

$$\begin{aligned} x(0) &= x_0, \\ \dot{x}(t) &= f(x(t)) + g(x(t))k(x(t)), \quad \forall t \in I(x_0). \end{aligned} \tag{2.12}$$

We are interested in designing controllers for (2.10) that guarantee the stability of the origin for the resulting closed-loop system by construction, which can be accomplished using a CLF. To this end, let $V : X \to \mathbb{R}_{\geq 0}$ be a Lyapunov function candidate and observe that the Lie derivative of V along the dynamics (2.10) is given by

$$\dot{V}(x, u) = \frac{\partial V}{\partial x}(x)(f(x) + g(x)u) = L_f V(x) + L_g V(x)u.$$

Here, $L_g V(x) = \frac{\partial V}{\partial x}(x)g(x) \in \mathbb{R}^{1 \times m}$ is the $1 \times m$ matrix whose components are the Lie derivatives of V along each column of g. At this point, we cannot ask if V is a Lyapunov function for (2.10)–this would first require fixing a controller $u = k(x)$, thereby producing a new dynamical system $\dot{x} = f_{\mathrm{cl}}(x)$, and then checking if $L_{f_{\mathrm{cl}}} V(x)$ satisfies the criteria of Definition 2.7. However, we can ask if it is possible, for each nonzero x, to pick some input u to enforce the Lyapunov conditions upon $\dot{V}(x, u)$. If V satisfies such a condition, then we say that V is a CLF. Formally, we define a CLF as follows.

Definition 2.8 (*Control Lyapunov function*) A Lyapunov function candidate $V : X \to \mathbb{R}_{\geq 0}$ is said to be a *control Lyapunov function* (CLF) for (2.10) if there exists $\alpha \in \mathcal{K}$ such that for all $x \in X \setminus \{0\}$

$$\inf_{u \in \mathcal{U}} \{L_f V(x) + L_g V(x)u\} < -\alpha(\|x\|).$$ (2.13)

Determining if a Lyapunov function candidate is a CLF depends heavily on the behavior of $L_g V$. For example, when $\mathcal{U} = \mathbb{R}^m$ (i.e., the control input is unconstrained) the condition in (2.13) can be restated as

$$\forall x \in X \setminus \{0\} : L_g V(x) = 0 \implies L_f V(x) < -\alpha(\|x\|).$$ (2.14)

That is, when $L_g V(x) \neq 0$ and the control input is unconstrained, it is always possible to pick some u such that the scalar inequality in (2.13) is satisfied; when $L_g V(x) = 0$ one must rely on the drift dynamics f to ensure the such a condition is met. If control constraints are present–for example, when $\mathcal{U} \subset \mathbb{R}^m$ is a convex polytope–determining if V is a CLF is a much more challenging problem, which will be discussed later in this section.

The appeal of a CLF is that a CLF induces an entire family of stabilizing policies expressed through the set-valued map

$$K_{\text{clf}}(x) := \{u \in \mathcal{U} \mid L_f V(x) + L_g V(x)u \leq -\alpha(\|x\|)\},$$ (2.15)

that assigns to each $x \in X$, a set $K_{\text{clf}}(x) \subset \mathcal{U}$ of control values satisfying the CLF condition from (2.13). The main result with regard to CLFs is that choosing any locally Lipschitz controller $u = k(x)$ satisfying $k(x) \in K_{\text{clf}}(x)$ for all $x \in X$ renders the origin asymptotically stable.

Theorem 2.4 *If V is a CLF for (2.10), then any locally Lipschitz controller $u = k(x)$ satisfying $k(x) \in K_{\text{clf}}(x)$ for all $x \in X$ renders the origin asymptotically stable.*

Proof Putting $u = k(x)$ and computing $\dot{V}$ reveals that

$$\dot{V}(x) = L_f V(x) + L_g V(x)k(x) \leq -\alpha(\|x\|),$$

which, according to Definition 2.7, implies that V is a Lyapunov function for the closed-loop system. From Theorem 2.2, it follows that the origin is asymptotically stable for the closed loop system. □

The notion of a CLF can also be specialized to handle exponential stabilization tasks using the notion of an *exponentially stabilizing CLF*.

Definition 2.9 (*Exponentially stabilizing CLF*) A continuously differentiable function $V : X \to \mathbb{R}_{\geq 0}$ is said to be an *exponentially stabilizing control Lyapunov function* (ES-CLF) if there exist positive constants $c_1, c_2, c_3 \in \mathbb{R}_{>0}$ such that for all $x \in X$

$$c_1 \|x\|^2 \leq V(x) \leq c_2 \|x\|^2,$$ (2.16)

and for all $x \in X \setminus \{0\}$

$$\inf_{u \in \mathcal{U}} \{L_f V(x) + L_g V(x)u\} < -c_3 \|x\|^2. \tag{2.17}$$

Similar to CLFs, an ES-CLF V induces a set-valued map $K_{es\text{-}clf} : X \rightrightarrows \mathcal{U}$ that associates to each $x \in X$ the set $K_{es\text{-}clf} \subset \mathcal{U}$ of control values satisfying the ES-CLF condition from (2.17) as

$$K_{clf}(x) := \{u \in \mathcal{U} \mid L_f V(x) + L_g V(x)u \leq -\alpha(\|x\|)\}. \tag{2.18}$$

Choosing any such controller $k(x) \in K_{es\text{-}clf}(x)$ for all $x \in X$ renders the closed-loop system ES-CLF as shown in the following theorem.

Theorem 2.5 *If V is an ES-CLF for (2.10), then any locally Lipschitz controller $u = k(x)$ satisfying $k(x) \in K_{es\text{-}clf}(x)$ for all $x \in X$ renders the origin exponentially stable.*

Proof With $u = k(x)$, the Lie derivative of V along the closed-loop dynamics satisfies

$$\dot{V}(x) = L_f V(x) + L_g V(x)k(x) \leq -c_3 \|x\|^2,$$

and the theorem follows from Theorem 2.3.   □

The existence of a CLF implies the existence of control inputs that, for each nonzero x, enforce negativity of $\dot{V}(x, u)$, and Theorem 2.4 illustrates that if such inputs can be stitched together into a locally Lipschitz feedback control policy, then that policy renders the origin asymptotically stable.

The approach taken in this book is to view the CLF condition (2.13) as a *constraint* that must be satisfied by the control input u for each $x \in X$. When viewed as a function of u, such a constraint is affine and is therefore convex, which allows the computation of control inputs satisfying the CLF condition (2.13) to be computed, for any $x \in X$, by solving a convex optimization problem. For example, inputs satisfying the CLF condition can be computed as the solution to the optimization problem

$$k(x) = \arg\min_{u \in \mathcal{U}} \quad \frac{1}{2} \|u\|^2$$
$$\text{subject to} \quad L_f V(x) + L_g V(x)u \leq -\alpha(\|x\|), \tag{2.19}$$

which is a quadratic program (QP) for a given, fixed x, provided $\mathcal{U} = \mathbb{R}^m$ or $\mathcal{U}$ is a convex polytope. The controller in (2.19) returns, for any given x, the input $u = k(x)$ of minimum norm that satisfies $k(x) \in K_{clf}(x)$ and is often referred to as a *pointwise min-norm controller*. Clearly, such a controller satisfies $k(x) \in K_{clf}(x)$ for each $x \in X$ and therefore renders the origin asymptotically stable provided $x \mapsto k(x)$ is locally Lipschitz on X. However, the fact that control inputs are computed as the solution to an optimization problem in (2.19), as opposed to being output from a closed-form feedback law, raises concerns regarding the continuity and smoothness properties of the resulting controller. The remainder of this section is thus dedicated to establishing various properties of the controller in (2.19). In particular, we will study the more general QP-based controller

$$k(x) = \arg\min_{u \in \mathbb{R}^m} \quad \tfrac{1}{2}\|u\|^2 - k_0(x)^\top u$$
$$\text{subject to} \quad a(x) + b(x)^\top u \le 0, \tag{2.20}$$

so that such results can be directly extended to other QP-based controllers introduced throughout the book. The following result provides conditions that guarantee the QP-based controller (2.20) is locally Lipschitz.

Theorem 2.6 (Lipschitz continuity of QP Controllers) *Consider the QP-based controller* $k : X \to \mathbb{R}^m$ *defined on some open subset* $X \subset \mathbb{R}^n$ *and suppose that* $k_0 : X \to \mathbb{R}^m$, $a : X \to \mathbb{R}$, *and* $b : X \to \mathbb{R}^m$ *are locally Lipschitz on* X. *Further, assume that*

$$\forall x \in X : b(x) = 0 \implies a(x) < 0. \tag{2.21}$$

Then, the solution to (2.20) can be expressed as

$$k(x) = \begin{cases} k_0(x), & \text{if } a(x) + b(x)^\top k_0(x) \le 0 \\ k_0(x) - \frac{a(x)+b(x)^\top k_0(x)}{\|b(x)\|^2} b(x), & \text{if } a(x) + b(x)^\top k_0(x) > 0, \end{cases} \tag{2.22}$$

and is locally Lipschitz on X.

Proof The proof leverages the Karush-Kuhn-Tucker (KKT) conditions for optimality. We first note that since the objective function is convex and differentiable, and the constraints are affine, the KKT conditions are necessary and sufficient for optimality. We next define the Lagrangian

$$L(x, u, \lambda) := \tfrac{1}{2}\|u\|^2 - k_0(x)^\top u + \lambda(a(x) + b(x)^\top u), \tag{2.23}$$

where $\lambda \in \mathbb{R}$ is the Lagrange multiplier. The KKT conditions state that a pair $(u^*(x), \lambda^*(x))$ is optimal if and only if the following conditions are satisfied:

$$\begin{aligned} \tfrac{\partial L}{\partial u}(x, u^*(x), \lambda^*(x)) &= 0 & \text{(stationarity)} \\ a(x) + b(x)^\top u^*(x) &\le 0 & \text{(primal feasibility)} \\ \lambda^*(x) &\ge 0 & \text{(dual feasibility)} \\ \lambda^*(x)(a(x) + b(x)^\top u^*(x)) &= 0 & \text{(complementary slackness)}. \end{aligned}$$

Hence, for (2.20) the KKT conditions imply that an optimal solution $(u^*(x), \lambda^*(x))$ must satisfy

$$u^*(x) = k_0(x) - \lambda^*(x)b(x). \tag{2.24}$$

We will derive the closed-form solution to (2.20) by breaking the dual feasibility condition down into two cases: (i) $\lambda^*(x) = 0$ and (ii) $\lambda^*(x) > 0$. If $\lambda^*(x) = 0$ then (2.24) implies that $u^*(x) = k_0(x)$. To determine the subset of the state space where this solution applies, we leverage the primal feasibility condition to see that if $u^*(x) = k_0(x)$ then $a(x) + b(x)^\top k_0(x) \leq 0$. This implies that $u^*(x) = k_0(x)$ is the optimal solution to the QP (2.20) in the set

$$\Omega_1 := \{x \in X \mid a(x) + b(x)^\top k_0(x) \leq 0\}. \tag{2.25}$$

We now show that the condition in (2.21) implies that any point such that $b(x) = 0$ lies strictly in the interior of Ω_1. Indeed, note from (2.21) that

$$b(x) = 0 \implies a(x) < 0 \implies a(x) + b(x)^\top k_0(x) < 0 \implies x \in \text{Int}(\Omega_1). \tag{2.26}$$

We now consider the case when $\lambda^*(x) > 0$. In such a case it follows from the complementary slackness condition that we must have

$$\begin{aligned}
0 &= a(x) + b(x)^\top u^*(x) \\
&= a(x) + b(x)^\top (k_0(x) - \lambda^*(x)b(x)) \\
&= a(x) + b(x)^\top k_0(x) - \lambda^*(x)\|b(x)\|^2,
\end{aligned} \tag{2.27}$$

where the second equality follows from substituting in the stationarity condition (2.24). To solve the above equation for λ^* we note that when $\lambda^*(x) > 0$ we must have $b(x) \neq 0$. Indeed, if it were not, then $b(x) = 0$, which would imply that $u^*(x) = k_0(x)$ and that $\lambda^*(x) = 0$, which contradicts the initial assumption that $\lambda^*(x) > 0$. Hence, using the fact that $b(x) \neq 0$ and solving for λ^* yields

$$\lambda^*(x) = \frac{a(x) + b(x)^\top k_0(x)}{\|b(x)\|^2}, \tag{2.28}$$

which, after substituting back into (2.24), yields

$$u^*(x) = k_0(x) - \frac{a(x) + b(x)^\top k_0(x)}{\|b(x)\|^2}b(x). \tag{2.29}$$

To determine the subset of the state space where the above solution applies, we note from the complementary slackness condition (2.27) that

$$a(x) + b(x)^\top k_0(x) = \lambda^*(x)\|b(x)\|^2 > 0 \tag{2.30}$$

where the inequality follows from $\lambda^*(x) > 0$ and $b(x) \neq 0$. Hence, the optimal solution to the QP (2.20) is given by (2.29) in the set

$$\Omega_2 := \{x \in X \mid a(x) + b(x)^\top k_0(x) > 0\}. \tag{2.31}$$

Combing the results for Ω_1 and Ω_2, the solution to (2.20) is given by

$$k(x) = \begin{cases} k_0(x) & \text{if } x \in \Omega_1, \\ k_0(x) - \frac{a(x)+b(x)^\top k_0(x)}{\|b(x)\|^2} b(x) & \text{if } x \in \Omega_2, \end{cases} \tag{2.32}$$

which coincides with the solution given in (2.22). Clearly, $x \mapsto k(x)$ is locally Lipschitz on Ω_1 as $x \mapsto k_0(x)$ is locally Lipschitz on $\mathcal{X} \supset \Omega_1$. Furthermore, $x \mapsto k(x)$ is locally Lipschitz on Ω_2 since $b(x) \neq 0$ for all $x \in \Omega_2$ and $x \mapsto k_0(x), a(x), b(x)$ are all locally Lipschitz on $\mathcal{X} \supset \Omega_2$. Given that k is locally Lipschitz on Ω_1 and Ω_2, it remains to show that k is continuous on

$$\bar{\Omega}_1 \cap \bar{\Omega}_2 = \partial \Omega_1 = \{x \in \mathcal{X} \mid a(x) + b(x)^\top k_0(x) = 0\}. \tag{2.33}$$

To this end, let $\{x_i\}_{i \in \mathbb{N}}$ be a convergent sequence in Ω_1 such that $\lim_{i \to \infty} x_i = x \in \partial \Omega_1$. Using the fact that $x \mapsto k_0(x)$ is continuous, we then have

$$\lim_{i \to \infty} k(x_i) = \lim_{i \to \infty} k_0(x_i) = k_0(x). \tag{2.34}$$

Now let $\{x_i\}_{i \in \mathbb{N}}$ be a convergent sequence in Ω_2 such that $\lim_{i \to \infty} x_i = x \in \partial \Omega_1$. Using the fact that $x \mapsto k_0(x)$, $x \mapsto a(x)$, and $x \mapsto b(x)$ are all continuous we have

$$\lim_{i \to \infty} k(x_i) = \lim_{i \to \infty} \left[k_0(x_i) - \frac{a(x_i)+b(x_i)^\top k_0(x_i)}{\|b(x_i)\|^2} b(x_i) \right] = k_0(x), \tag{2.35}$$

where the last equality follows from $x \in \partial \Omega_1$ and (2.33). Note that the limit in (2.35) exists because $\lim_{i \to \infty} b(x_i) \neq 0$ for the sequence $\{x_i\}_{i \in \mathbb{N}}$ in Ω_2 converging to $x \in \partial \Omega_1$ since the set of points where $b(x) = 0$ lies strictly in the interior of Ω_1. Combining (2.34) and (2.35) reveals that $x \mapsto k(x)$ is continuous on $\bar{\Omega}_1 \cap \bar{\Omega}_2$, which, along with the fact that $x \mapsto k(x)$ is locally Lipschitz on Ω_1 and Ω_2, reveals that $x \mapsto k(x)$ is locally Lipschitz on $\mathcal{X}$. $\square$

Remark 2.1 Note that in the above proof, establishing Lipschitz continuity of the QP-based controller relies heavily on the fact that a strict, rather than a nonstrict, inequality is used in (2.21). If a nonstrict inequality is used then it is possible that b may vanish on $\partial \Omega_1$ from (2.33), in which case the limit in (2.35) may not exist.

A straightforward application of the preceding theorem reveals that the solution to the CLF-QP (2.19) when $\mathcal{U} = \mathbb{R}^m$ is given by

$$k(x) = \begin{cases} 0 & \text{if } L_f V(x) + \alpha(\|x\|) \leq 0, \\ -\frac{L_f V(x)+\alpha(\|x\|)}{\|L_g V(x)^\top\|^2} L_g V(x)^\top & \text{if } L_f V(x) + \alpha(\|x\|) > 0, \end{cases} \tag{2.36}$$

and is locally Lipschitz on $\mathcal{X} \setminus \{0\}$. In general, the above controller may fail to be continuous at the origin. However, such a controller can be made to be continuous at the origin (i.e., we may take $k(0) = 0$ while preserving continuity) provided the corresponding CLF satisfies the following condition.

Definition 2.10 A CLF V is said to satisfy the *small control property* if for each $\varepsilon \in \mathbb{R}_{>0}$ there exists a $\delta \in \mathbb{R}_{>0}$ such that if $x \in \mathcal{B}_\varepsilon(0) \setminus \{0\}$, then there exists a $u \in \mathcal{B}_\delta(0)$ such that $L_f V(x) + L_g V(x) u \le -\alpha(\|x\|)$.

The small control property guarantees that for states arbitrarily close to the origin, there exists small enough control inputs that enforce negativity of $\dot{V}$.

The QP-based approach to control brings with it the ability to incorporate additional objectives into the control design, expressed as additional constraints in the QP. For example, polytopic control constraints can be incorporated by simply including the additional halfspace constraint $A_0 u \le b_0$, where A_0 and b_0 are a matrix and vector, respectively, of appropriate dimensions describing the control constraint set $\mathcal{U} = \{u \in \mathbb{R}^m \mid A_0 u \le b_0\}$. Although it is straightforward in practice to incorporate such a constraint, care must be taken to ensure feasibility of the QP and satisfaction of the ultimate control objective. For example, if it cannot be verified that V is a valid CLF under input constraints, then there may not exist control values simultaneously satisfying the input constraints and the CLF condition, leading to infeasibility of the QP. Even if simultaneous satisfaction of such constraints can be guaranteed, there are few results that ensure the resulting controller will be sufficiently smooth to guarantee existence and uniqueness of solutions to the closed-loop system. One way to address this challenge is to relax a subset of constraints and then penalize the magnitude of this relaxation in the objective function. Since control constraints are typically an intrinsic property of the system under consideration and cannot be relaxed, one can relax the CLF constraint with a scalar relaxation variable $\delta \in \mathbb{R}$ and solve a relaxed version of the original CLF-QP from (2.19) as

$$
k(x), \delta^* = \arg\min_{u \in \mathcal{U}, \, \delta \in \mathbb{R}} \quad \frac{1}{2}\|u\|^2 + p\delta^2
$$
$$
\text{subject to} \quad L_f V(x) + L_g V(x) u \le -\alpha(\|x\|) + \delta, \tag{2.37}
$$

where $p \in \mathbb{R}_{>0}$ is a relaxation penalty. The above controller respects the actuation limitations of the system but does not necessarily guarantee stability. Despite this theoretical limitation, the relaxed CLF-QP controller in (2.37) has been successfully used in practice (see the notes in Sect. 2.4). The idea of relaxing constraints in an effort to balance potentially competing control objectives will be further explored in Chap. 3 in the context of safety-critical control.

2.3   Designing Control Lyapunov Functions

The benefits of the CLF approach discussed thus far are contingent upon the construction of a valid CLF. In this section, we present two approaches–feedback linearization and backstepping–for systematically constructing CLFs for special classes of nonlinear control systems. We keep our exposition brief, only focusing on the main concepts. Further discussions and references to technical details can be found in Sect. 2.4.

2.3.1 Feedback Linearization

In this section, we briefly outline the technique of feedback linearization as a means to systematically construct CLFs for certain classes of nonlinear control systems. The main idea behind feedback linearization is, as the name suggests, to transform a nonlinear system into a linear one by means of feedback, allowing linear control tools to be leveraged to complete the control design. The benefit of combing feedback linearization with CLFs is that once a feedback linearization-based controller is known, converse Lyapunov theorems can be invoked to produce a Lyapunov function certifying the stability of the control design, which, by definition, is also a CLF for the original nonlinear system. Hence, rather than implementing the controller that fully cancels the nonlinearities of the system, one can use any controller satisfying the CLF conditions, such as the QP-based controller proposed in Eq. (2.19).

We focus on the nonlinear control system (2.10), restated here for convenience:

$$\dot{x} = f(x) + g(x)u,$$

to which we associate an *output*

$$y = h(x), \tag{2.38}$$

where $h : X \to \mathbb{R}^m$ maps each element of the state space $X \subset \mathbb{R}^n$ to a vector of outputs $y \in \mathbb{R}^m$. Note that the dimension of the output is the same as that of the control input. Our objective is to design a controller for (2.10) such that the output is driven to zero. Accomplishing this objective depends heavily on the notion of the *relative degree* of the output with respect to the system dynamics. Informally, the relative degree of a component of the output (2.38) with respect to system (2.10) is the number of times that output component needs to be differentiated along the system dynamics for the control input to explicitly appear. Obviously, different output components can have different relative degrees. The collection of the relative degrees for each output component forms the vector relative degree of the output. A more detailed discussion and references can be found in Sect. 2.4. Before proceeding we introduce the notion of higher order Lie derivatives obtained by taking the Lie derivative of a Lie derivative. For example, given the output $h : X \to \mathbb{R}^m$ and the vector field $f : X \to \mathbb{R}^n$, the second order Lie derivative of h along f is defined as

$$L_f^2 h(x) = \frac{\partial L_f h}{\partial x}(x) f(x).$$

In general, we denote the r-th Lie derivative as

$$L_f^r h(x) = \frac{\partial L_f^{r-1} h}{\partial x}(x) f(x).$$

It is also possible to take higher order Lie derivatives along different vector fields. For example, we can take the Lie derivative of $L_f h$ along g as

$$L_g L_f h(x) = \frac{\partial L_f h}{\partial x}(x) g(x).$$

For the remainder of this section, we assume that the vector relative degree of the output is well-defined (i.e., all components have the same relative degree) on X with respect to (2.10), and it is equal to 2. Under this assumption, differentiation of the output along the system dynamics yields

$$\begin{aligned} \dot{y} &= L_f h(x) \\ \ddot{y} &= L_f^2 h(x) + L_g L_f h(x) u, \end{aligned} \tag{2.39}$$

where $L_g L_f h : X \to \mathbb{R}^{m \times m}$ is referred to as the *decoupling matrix*, which is invertible on X provided the relative degree 2 condition holds. Applying the control

$$u = (L_g L_f h(x))^{-1} \left(-L_f^2 h(x) + \mu \right), \tag{2.40}$$

where $\mu \in \mathbb{R}^m$ is an auxiliary input to be specified and defining the coordinates $\eta := [y^\top \; \dot{y}^\top]^\top \in \mathbb{R}^{2m}$ yields the linear system

$$\dot{\eta} = F\eta + G\mu, \tag{2.41}$$

with

$$F = \begin{bmatrix} 0_{m \times m} & I_{m \times m} \\ 0_{m \times m} & 0_{m \times m} \end{bmatrix}, \quad G = \begin{bmatrix} 0_{m \times m} \\ I_{m \times m} \end{bmatrix}, \tag{2.42}$$

where $0_{m \times m} \in \mathbb{R}^{m \times m}$ is an $m \times m$ matrix of zeros. Choosing the auxiliary input as $\mu = -K\eta$, such that the closed-loop system matrix $A := F - GK$ is Hurwitz renders the origin of $\dot{\eta} = A\eta$ exponentially stable. From standard converse Lyapunov theorems, it follows that, for any symmetric positive definite $Q \in \mathbb{R}^{2m \times 2m}$, there exists a symmetric positive definite $P \in \mathbb{R}^{2m \times 2m}$ solving the Lyapunov equation

$$PA + A^\top P = -Q, \tag{2.43}$$

such that

$$V(\eta) = \eta^\top P \eta \tag{2.44}$$

is a Lyapunov function certifying exponential stability of the closed-loop system's origin. In particular, taking the Lie derivative of V along the closed-loop vector field and leveraging the Lyapunov equation (2.43) leads to

$$\dot{V}(\eta) = 2\eta^\top P A \eta = \eta^\top (PA + A^\top P)\eta = \eta^\top Q\eta \leq -\lambda_{\min}(Q) \|\eta\|^2. \tag{2.45}$$

Hence, V is also an ES-CLF for the output dynamics as V satisfies

$$\lambda_{\min}(P) \|\eta\|^2 \leq V(\eta) \leq \lambda_{\max}(P) \|\eta\|^2, \quad \forall \eta \in \mathbb{R}^{2m}, \tag{2.46a}$$

$$\inf_{u \in \mathbb{R}^m} \dot{V}(\eta, u) < -c\lambda_{\min}(Q)\|\eta\|^2, \quad \forall \eta \in \mathbb{R}^{2m} \setminus \{0\}, \tag{2.46b}$$

with $c \in (0, 1)$, which can be used to generate a controller that exponentially drives η to zero by solving the QP

$$\min_{u \in \mathbb{R}^m} \quad \frac{1}{2}\|u\|^2$$
$$\text{subject to} \quad \dot{V}(\eta, u) \leq -c\lambda_{\min}(Q)\|\eta\|^2. \tag{2.47}$$

2.3.2 Backstepping

In this section, we discuss backstepping as a means to systematically construct CLFs for a particular class of nonlinear control systems. Backstepping is a recursive design procedure applicable to classes of nonlinear control systems with a hierarchical structure, in which higher order states act as "virtual" control inputs for the lower order dynamics. This approach allows for systematically constructing a CLF for the overall system using only a CLF for the lowest order subsystem. To this end, consider a nonlinear control affine system in *strict feedback form*

$$\begin{aligned}\dot{x} &= f_0(x) + g_0(x)\xi \\ \dot{\xi} &= f_1(x, \xi) + g_1(x, \xi)u,\end{aligned} \tag{2.48}$$

where $(x, \xi) \in \mathbb{R}^n \times \mathbb{R}^p$ is the system state, $u \in \mathbb{R}^m$ is the control input, and the functions $f_0 : \mathbb{R}^n \to \mathbb{R}^n$, $g_0 : \mathbb{R}^n \to \mathbb{R}^{n \times p}$, $f_1 : \mathbb{R}^n \times \mathbb{R}^p \to \mathbb{R}^p$, $g_1 : \mathbb{R}^n \times \mathbb{R}^p \to \mathbb{R}^{p \times m}$ characterize the system dynamics. We assume that the functions characterizing the dynamics are locally Lipschitz and that g_1 is pseudo-invertible on its domain with $g_1(x, \xi)^\dagger := (g_1(x, \xi)^\top g_1(x, \xi))^{-1} g_1(x, \xi)^\top$ denoting the Moore-Penrose pseudo-inverse. Our main objective is to design a controller that renders $x = 0$ an asymptotically stable equilibrium point for the closed-loop system. The backstepping methodology proceeds by viewing ξ as a "virtual" control input for the subsystem

$$\dot{x} = f_0(x) + g_0(x)\xi, \tag{2.49}$$

and then designing a controller $k_0 : \mathbb{R}^n \to \mathbb{R}^p$ that would render the origin of the closed-loop system

$$\dot{x} = f_0(x) + g_0(x)k_0(x),$$

asymptotically stable, provided we could simply choose $\xi = k_0(x)$.

Let $V_0 : \mathbb{R}^n \to \mathbb{R}_{\geq 0}$ be a twice-continuously differentiable CLF for the first subsystem in the sense that there exists a twice-continuously differentiable controller $k_0 : \mathbb{R}^n \to \mathbb{R}^p$ and $\alpha_1, \alpha_2, \alpha_3 \in \mathcal{K}_\infty$ such that, for all $x \in \mathbb{R}^n$,

$$\alpha_1(\|x\|) \leq V_0(x) \leq \alpha_2(\|x\|),$$

$$L_{f_0} V_0(x) + L_{g_0} V_0(x) k_0(x) \leq -\alpha_3(\|x\|).$$

Now define the coordinate transformation $z := \xi - k_0(x)$, which represents the difference between ξ and the desired control we would implement on the first subsystem (2.49) if ξ were directly controllable. Using the $(x, z) \in \mathbb{R}^n \times \mathbb{R}^p$ coordinates, system (2.48) can be represented as

$$\begin{aligned}
\dot{x} &= f_0(x) + g_0(x) k_0(x) + g_0(x) z \\
\dot{z} &= f_1(x, \xi) + g_1(x, \xi) u - \frac{\partial k_0}{\partial x}(x)(f_0(x) + g_0(x) k_0(x) + g_0(x) z).
\end{aligned} \tag{2.50}$$

Now consider the composite Lyapunov function candidate

$$V(x, z) = V_0(x) + \tfrac{1}{2} z^\top z,$$

whose time derivative is

$$\begin{aligned}
\dot{V} &= L_{f_0} V_0(x) + L_{g_0} V_0(x) k_0(x) + L_{g_0} V_0(x) z \\
&\quad + z^\top \left(f_1(x, \xi) + g_1(x, \xi) u - \frac{\partial k_0}{\partial x}(x)(f_0(x) + g_0(x) k_0(x) + g_0(x) z) \right).
\end{aligned} \tag{2.51}$$

Choosing the control as

$$u = g_1(x, \xi)^\dagger \left(-f_1(x, \xi) + \frac{\partial k_0}{\partial x}(x)(f_0(x) + g_0(x) k_0(x) + g_0(x) z) - L_{g_0} V_0(x)^\top - K z \right), \tag{2.52}$$

where $K \in \mathbb{R}^{p \times p}$ is positive definite, yields

$$\begin{aligned}
\dot{V} &= L_{f_0} V_0(x) + L_{g_0} V_0(x) k_0(x) - z^\top K z \\
&\leq -\alpha_3(\|x\|) - \lambda_{\min}(K) \|z\|^2,
\end{aligned} \tag{2.53}$$

which implies that V is a Lyapunov function for the transformed system (2.50). Hence, V is also a CLF for the transformed system (2.50) as V satisfies

$$\alpha_1(\|x\|) + \tfrac{1}{2} \|z\|^2 \leq V(x, z) \leq \alpha_2(\|x\|) + \tfrac{1}{2} \|z\|^2, \ \forall (x, z) \in \mathbb{R}^n \times \mathbb{R}^p \tag{2.54a}$$

$$\inf_{u \in \mathbb{R}^m} \dot{V}(x, z, u) < -c(\alpha_3(\|x\|) + \lambda_{\min}(K) \|z\|^2), \ \forall (x, z) \in (\mathbb{R}^n \setminus \{0\}) \times (\mathbb{R}^p \setminus \{0\}). \tag{2.54b}$$

Although the preceding discussion focused on hierarchical systems with only 2 subsystems, the same steps can be recursively followed for systems with an arbitrary number $q \in \mathbb{N}$ of finite subsystems of the form

$$\dot{x} = f_0(x) + g_0(x)\xi_1$$
$$\dot{\xi}_1 = f_1(x, \xi_1) + g_1(x, \xi_1)\xi_2$$
$$\dot{\xi}_2 = f_2(x, \xi_1, \xi_2) + g_2(x, \xi_1, \xi_2)\xi_3$$
$$\vdots$$
$$\dot{\xi}_q = f_q(x, \xi_1, \xi_2, \ldots, \xi_q) + g_q(x, \xi_1, \xi_2, \ldots, \xi_q)u.$$

2.3.3   Design Example

We close this section with a simple example that demonstrates the construction of a CLF using the methods outlined thus far. We consider an inverted pendulum with angular position $q \in \mathbb{R}$ whose equations of motions can be expressed as

$$m\ell^2\ddot{q} - mg\ell \sin(q) = \tau - b\dot{q}, \tag{2.55}$$

where $m \in \mathbb{R}_{>0}$ is the pendulum's mass, assumed to be concentrated at its tip, $\ell \in \mathbb{R}_{>0}$ is the pendulum's length, $g \in \mathbb{R}_{>0}$ is the acceleration due to gravity, $b \in \mathbb{R}_{>0}$ is a viscous damping coefficient, and $\tau \in \mathbb{R}$ is the torque applied to the base of the pendulum. Taking the state of the system as $x = [q \; \dot{q}]^\top$ and the control input as $u = \tau$ allows the pendulum to be expressed in control-affine form (2.10) as:

$$\dot{x} = \underbrace{\begin{bmatrix} \dot{q} \\ \frac{g}{\ell}\sin(q) - \frac{b}{m\ell^2}\dot{q} \end{bmatrix}}_{f(x)} + \underbrace{\begin{bmatrix} 0 \\ \frac{1}{m\ell^2} \end{bmatrix}}_{g(x)} u. \tag{2.56}$$

Our main objective is to design a feedback controller $u = k(x)$ that drives the pendulum to $q = 0$ using a CLF, which we'll accomplish using both feedback linearization and backstepping.

Feedback Linearization

To proceed with the feedback linearization approach, we define the output $y = h(x) = q$ and compute the Lie derivative of y to get $\dot{y} = L_f h(x) = \dot{q}$. As the input u does not appear in $\dot{y}$, the relative degree of the output is greater than one, and we proceed by computing the Lie derivative of $L_f h$ to get

$$\ddot{y} = \ddot{q} = \underbrace{\frac{g}{\ell}\sin(q) - \frac{b}{m\ell^2}\dot{q}}_{L_f^2 h(x)} + \underbrace{\frac{1}{m\ell^2}}_{L_g L_f h(x)} u. \tag{2.57}$$

As $L_g L_f h(x) \neq 0$ for all $x \in X = \mathbb{R}^2$, the output has relative degree 2 on X. Applying the feedback

$$u = L_g L_f h(x)^{-1} \left(-L_f^2 h(x) + \mu \right) = m\ell^2 \left(-\frac{g}{\ell} \sin(q) + \frac{b}{m\ell^2} \dot{q} + \mu \right),$$

and defining $\eta = [y \; \dot{y}]^\top$ yields the linear control system

$$\dot{\eta} = \begin{bmatrix} 0 & 1 \\ 0 & 0 \end{bmatrix} \eta + \begin{bmatrix} 0 \\ 1 \end{bmatrix} \mu.$$

Using the proportional-derivative controller $\mu = -K_p y - K_d \dot{y}$, with $K_p, K_d \in \mathbb{R}_{>0}$ such that the origin of the closed-loop linear system

$$\dot{\eta} = \underbrace{\begin{bmatrix} 0 & 1 \\ -K_p & -K_d \end{bmatrix}}_{A} \eta,$$

is exponentially stable, allows for constructing a Lyapunov function $V(\eta) = \eta^\top P \eta$, where $P \in \mathbb{R}^{2\times2}$ is the positive definite solution to the Lyapunov equation (2.43) for any positive definite $Q \in \mathbb{R}^{2\times2}$. For this particularly simple output, we have $\eta = [y \; \dot{y}]^\top = [q \; \dot{q}]^\top = x$. Thus converting back to the original coordinates x, the Lyapunov function for the original system is given by $V(x) = x^\top P x$, which is, by definition, a CLF for the original nonlinear control system.

Backstepping

To proceed with the backstepping approach, we first represent the system in strict-feedback form with state $(x, \xi) = (q, \dot{q})$ and control $u = \tau$ as

$$\begin{aligned} \dot{x} &= \xi \\ \dot{\xi} &= \frac{g}{\ell} \sin(x) - \frac{b}{m\ell^2} \xi + \frac{1}{m\ell^2} u. \end{aligned} \tag{2.58}$$

We then design the "virtual" controller for the first subsystem $k_0(x) = -K_p x$ with $K_p \in \mathbb{R}_{>0}$, whose stability can be certified using the Lyapunov function $V_0(x) = \frac{1}{2} x^2$. This function is twice-continuously differentiable and satisfies the Lyapunov conditions with $\alpha_1(s) = \alpha_2(s) = \frac{1}{2} s^2$ and $\alpha_3(s) = K_p s^2$. We next introduce the coordinate transformation $z = \xi - k_0(x) = \xi + K_p x$ with dynamics

$$\begin{aligned} \dot{z} &= \dot{\xi} - \frac{\partial k_0}{\partial x}(x) \dot{x} \\ &= \frac{g}{\ell} \sin(x) - \frac{b}{m\ell^2} \xi + \frac{1}{m\ell^2} u + K_p \xi. \end{aligned} \tag{2.59}$$

Thus, the dynamics in the (x, z) coordinates can be expressed as

$$\dot{x} = z - K_p x$$

$$\dot{z} = \frac{g}{\ell} \sin(x) - \frac{b}{m\ell^2}\xi + \frac{1}{m\ell^2}u + K_p z - K_p^2 x. \tag{2.60}$$

Now consider the Lyapunov function candidate $V(x, z) = \frac{1}{2}x^2 + \frac{1}{2}z^2$, whose time-derivative is

$$\dot{V} = x\left[z - K_p x\right] + z\left[\frac{g}{\ell}\sin(x) - \frac{b}{m\ell^2}\xi + \frac{1}{m\ell^2}u + K_p z - K_p^2 x\right] \tag{2.61}$$

Taking the control as

$$u = m\ell^2\left[-\frac{g}{\ell}\sin(x) + \frac{b}{m\ell^2}\xi - K_p z + K_p^2 x - x - K_d z\right], \tag{2.62}$$

with $K_d \in \mathbb{R}_{>0}$ then yields

$$\dot{V} = -K_p x^2 - K_d z^2. \tag{2.63}$$

Hence, V is a CLF for the original nonlinear control system as satisfies the conditions from (2.54) with $\alpha_1(s) = \alpha_2(s) = \frac{1}{2}s^2$, $\alpha_3(s) = K_p s^2$, and $K = K_d$.

2.4 Notes

In this chapter we introduced background on nonlinear dynamical systems as well as the fundamentals of Lyapunov theory in both the analysis and control of nonlinear systems. For a more in-depth treatment of ordinary differential equations we refer the reader to [1, 2], whereas more details on nonlinear systems analysis can be found in [3–6]. The standard definitions for stability of an equilibrium point are typically given an $\epsilon - \delta$ characterization. In this book we have opted instead to leverage comparison functions in Sect. 2.1.1, which provide a concise and elegant framework for characterizing stability. Our choice to use comparison functions is also motivated by the characterizations of safety presented in the following chapter and its duality with stability. One of the earliest works that leverages comparison functions to define notions of stability is the book by Hahn [7] and the work by Sontag in [8] made the use of comparison functions more common in the control-theoretic literature. A survey of comparison functions, including a more complete history and various technical results, can be found in [9].

The notion of Control Lyapunov Function (CLF) discussed in Sect. 2.2 was first introduced by Artstein [10] in 1983, where it was shown that the existence of a continuously differentiable CLF is sufficient for the existence of a controller, locally Lipschitz everywhere except possibly the origin, that asymptotically stabilizes the origin of the closed-loop system. The results in [10], however, are non-constructive in the sense that no explicit formula for constructing such a controller from a CLF is provided. In 1989, Sontag [11] provided a "universal" formula using what is now commonly referred to as Sontag's Formula to explicitly construct a controller from a given CLF. Other universal constructions of controllers from CLFs were proposed in [12], which introduced the point-wise min-norm controller that selects, at each state, the control value of minimum norm satisfying the CLF conditions. It was shown in [12] that such controllers are *inverse optimal* in the sense that any point-wise min-norm controller is also the optimal controller for a meaningful optimal control problem. Such controllers have shown a large amount of practical success in controlling complex nonlinear systems, such as bipedal robots [13].

Although it was clear in [12] that the point-wise min-norm controller is the solution to a convex optimization problem (2.19), namely a quadratic program (QP), explicitly solving these optimization problems to generate the corresponding control actions only became popular in the last decade [14–19]. The shift from traditional closed-form feedback controllers to optimization-based feedback controllers has been motivated by the ability to incorporate multiple, potentially competing, objectives into a single controller by carefully selecting the cost function and constraints in the resulting optimization problem. For example, the authors of [15] impose torque limits on the CLF controller for bipedal robots from [13] by embedding both the CLF conditions and torque limits as constraints in a QP, resulting in the relaxed CLF-QP from (2.37). Another example motivating this shift is the ability to combine CLFs with the *control barrier functions* discussed in the next chapter to balance performance and safety objectives.

The Karush-Kuhn-Tucker (KKT) conditions used to derive the closed-form solution to the CLF-QP (2.19) can be found in popular books on convex optimization, see, for example, that of Boyd and Vandenberghe [20] or that of Bertsekas [21]. Results establishing continuity and/or smoothness of more general optimization-based controllers can be found in [16, 22, 23]. Our proof of Theorem 2.6 is inspired by those in [24, 25]. More details on the small control property for ensuring continuity of the CLF-QP controller at the origin can be found in [23, 26].

The feedback linearization technique described in Sect. 2.3.1 has been a popular tool for nonlinear control design since its inception, which can be traced back to the work of Brockett [27]. For an in-depth treatment of feedback linearization, we refer the reader to books on geometric nonlinear control [28, 29]. Traditionally, a criticism of the feedback linearization approach is that it may cancel useful nonlinearities, such as damping terms that could otherwise aid the stabilization objective. More recently, works such as [13] have demonstrated the effectiveness of this methodology as a means to generate CLFs for nonlinear control systems that are feedback equivalent to linear systems. Our discussion on the connections between feedback linearization and CLFs follows that of [13].

Backstepping, reviewed in Sect. 2.3.2, was introduced in the early 1990s as an alternative to feedback linearization-based designs. By recursively constructing a CLF by "backstepping" through each subsystem, such designs are often able to avoid the cancellation of useful nonlinearities that may be cancelled by feedback linearization-based techniques. For a more in-depth treatment of backstepping we refer the reader to [26], and for a more complete survey of nonlinear control designs we refer the reader to [30]. Our discussion on the connections between backstepping and CLFs is inspired by that in [31].

References

1. Arnold VI (1978) Ordinary differential equations. MIT Press
2. Hirsch MW, Smale S (1974) Differential equations, dynamical systems, and linear algebra. Academic Press
3. Khalil HK (2002) Nonlinear systems, 3rd ed. Prentice Hall
4. Sontag ED (2013) Mathematical control theory: deterministic finite dimensional systems. Springer Science & Business Media
5. Slotine JJE, Li W (1991) Applied nonlinear control. Prentice Hall
6. Haddad WM, Chellaboina VS (2011) Nonlinear dynamical systems and control: a lyapunov-based approach. Princeton University Press
7. Hahn W (1967) Stability of motion. Springer
8. Sontag ED (1989) Smooth stabilization implies coprime factorization. IEEE Trans Autom Control 34(4):435–443
9. Kellett CM (2014) A compendium of comparison function results. Math Control Signals Syst 26:339–374
10. Artstein Z (1983) Stabilization with relaxed controls. Nonlinear Anal Theory Methods Appl 7(11):1163–1173
11. Sontag ED (1989) A universal construction of artstein's theorem on nonlinear stabilization. Syst Control Lett 13:117–123
12. Freeman RA, Kokotovic PV (1996) Inverse optimality in robust stabilization. SIAM J Control Optim 34(4):1365–1391
13. Ames AD, Galloway K, Sreenath K, Grizzle JW (2014) Rapidly exponentially stabilizing control lyapunov functions and hybrid zero dynamics. IEEE Trans Autom Control 59(4):876–891
14. Ames AD, Powell M (2013) Towards the unification of locomotion and manipulation through control lyapunov functions and quadratic programs. Control of cyber-physical systems, pp 219–240
15. Galloway K, Sreenath K, Ames AD, Grizzle JW (2015) Torque saturation in bipedal robotic walking through control lyapunov function-based quadratic programs. IEEE Access, vol 3
16. Morris BJ, Powell MJ, Ames AD (2015) Continuity and smoothness properties of nonlinear optimization-based feedback controllers. In: Proceedings of the IEEE conference on decision and control, pp 151–158
17. Nguyen Q, Sreenath K (2015) Optimal robust control for bipedal robots through control lyapunov function based quadratic programs. In: Robotics: science and systems
18. Ames AD, Grizzle JW, Tabuada P (2014) Control barrier function based quadratic programs with application to adaptive cruise control. In: Proceedings of the IEEE conference on decision and control, pp 6271–6278
19. Ames AD, Xu X, Grizzle JW, Tabuada P (2017) Control barrier function based quadratic programs for safety critical systems. IEEE Trans Autom Control 62(8):3861–3876
20. Boyd S, Vandenberghe L (2004) Convex optimization. Cambridge University Press
21. Bertsekas DP (2016) Nonlinear programming, 3rd ed. Athena Scientific
22. Hager WH (1979) Lipschitz continuity for constrained processes. SIAM J Control Optim 17(3):321–338

23. Jankovic M (2018) Robust control barrier functions for constrained stabilization of nonlinear systems. Automatica 96:359–367
24. Molnar TG, Kiss AK, Ames AD, Orosz G (2022) Safety-critical control with input delay in dynamic environment. IEEE Trans Control Syst Technol
25. Tan X, Cortez WS, Dimarogonas DV (2022) High-order barrier functions: robustness, safety and performance-critical control. IEEE Trans Autom Control 67(6):3021–3028
26. Krstić M, Kanellakopoulos I, Kokotović P (1995) Nonlinear and adaptive control design. Wiley
27. Brockett R (1978) Feedback invariants for nonlinear systems. IFAC Proc Vol 11(1):1115–1120
28. Isidori A (1995) Nonlinear control systems, 3rd ed. Springer
29. Nijmeijer H, van der Schaft A (2015) Nonlinear dynamical control systems. Springer
30. Kokotovic PV, Arcak M (2001) Constructive nonlinear control: a historical perspective. Automatica 37:637–662
31. Taylor AJ, Ong P, Molnar TG, Ames AD (2022) Safe backstepping with control barrier functions. In: Proceedings of the IEEE conference on decision and control, pp 5775–5782

Safety-Critical Control

In the preceding chapter, we discussed the fundamentals of Lyapunov theory, and how these ideas can be used to design controllers enforcing stability of dynamical systems. In the present chapter, we discuss how such ideas can be transposed to address the problem of safety. Informally, safety can be thought of as requiring a system to never do anything "bad." This abstract notion is dependent on the application under consideration. For example, in autonomous driving, safety may correspond to an autonomous vehicle never leaving its current lane, whereas for a robot navigating in a cluttered environment, safety may correspond to avoiding collisions with obstacles. In this book, the notion of safety is linked to set invariance, in the sense that a system is safe if it never leaves a set deemed "good". In the last chapter of the book (Chap. 9), we will briefly discuss the satisfaction of temporal logic formulas, which includes (is strictly more expressive than) safety, and is usually referred to as "correctness". Safety is defined in Sect. 3.1, before control barrier functions used to enforce it are introduced in Sects. 3.2 and 3.3. Final remarks, references, and suggestions for further reading are included in Sect. 3.4.

3.1 Safety and Set Invariance

In this section, we provide a definition of safety for dynamical systems and provide conditions under which dynamical systems can be guaranteed to be safe. Formally, the notion of safety is linked with the fundamental concept of *set invariance*.

Definition 3.1 (*Forward invariance and safety*) A set $C \subset \mathbb{R}^n$ is said to be *forward invariant* for (2.1) if, for each $x_0 \in C$, the trajectory $x : I(x_0) \to \mathbb{R}^n$ with $x(0) = x_0$ satisfies $x(t) \in C$, for all $t \in I(x_0)$. If C is forward invariant, then the system is said to be *safe* on C.

© The Author(s), under exclusive license to Springer Nature Switzerland AG 2023
M. Cohen and C. Belta, *Adaptive and Learning-Based Control of Safety-Critical Systems*, Synthesis Lectures on Computer Science, https://doi.org/10.1007/978-3-031-29310-8_3

Similar to the notions of stability from Definition 2.5, the above definition of safety requires knowledge of the system trajectory, which motivates the development of conditions imposed on the system vector field f that can be used to certify safety. For general closed subsets $C \subset \mathbb{R}^n$, such a development relies on the concept of *tangent cone* to a set.

Definition 3.2 For a closed set $C \subset \mathbb{R}^n$, the *Bouligand tangent cone*[1] to C at a point $x \in C$ is defined as

$$\mathcal{T}_C(x) := \left\{ v \in \mathbb{R}^n \,\middle|\, \liminf_{\tau \to 0^+} \frac{\|x + \tau v\|_C}{\tau} = 0 \right\}, \tag{3.1}$$

where

$$\|x\|_C := \inf_{y \in C} \|x - y\|$$

is the distance from a point $x \in \mathbb{R}^n$ to the set C.

Note that, if $x \in \text{Int}(C)$, then $\mathcal{T}_C(x) = \mathbb{R}^n$, and if $x \notin C$, then $\mathcal{T}_C(x) = \emptyset$. It is only on the boundary of C that $\mathcal{T}_C(x)$ becomes interesting. Informally, for $x \in \partial C$ the Bouligand tangent cone $\mathcal{T}_C(x)$ is the set of all vectors $v \in \mathbb{R}^n$ that are tangent to or point into C. The following result provides necessary and sufficient conditions for the forward invariance of closed sets for dynamical systems with a locally Lipschitz vector field.

Theorem 3.1 (Nagumo's Theorem) *A closed set $C \subset \mathbb{R}^n$ is forward invariant for (2.1) if and only if for all $x \in \partial C$*

$$f(x) \in \mathcal{T}_C(x). \tag{3.2}$$

The above result, often referred to as *Nagumo's Theorem* or *Brezis's Theorem*, simply states that C is forward invariant if and only if, for each point on the boundary of C, the vector field f at such points is either tangent to or points into C. A challenge with directly applying Theorem 3.1 is that for a general closed set $C \subset \mathbb{R}^n$, the computation of the tangent cone may be nontrivial. To provide more practical conditions to certifying safety, we now specialize the class of sets whose forward invariance we wish to certify. In particular, we consider sets $C \subset \mathbb{R}^n$ that can be expressed as the zero superlevel set of a continuously differentiable function $h : \mathbb{R}^n \to \mathbb{R}$ as

$$\begin{aligned}
C &= \{x \in \mathbb{R}^n \mid h(x) \geq 0\}, \\
\partial C &= \{x \in \mathbb{R}^n \mid h(x) = 0\}, \\
\text{Int}(C) &= \{x \in \mathbb{R}^n \mid h(x) > 0\}.
\end{aligned} \tag{3.3}$$

---

[1] The Bouligand tangent cone is also referred to at the contingent cone.

For sets of the form (3.3), it can be shown that, provided[2] that $\nabla h(x) \neq 0$ for all $x \in \partial C$, the tangent cone to C at $x \in \partial C$ can be represented as:

$$\mathcal{T}_C(x) = \{v \in \mathbb{R}^n \mid \nabla h(x)^\top v \geq 0\}. \tag{3.4}$$

A straightforward application of Theorem 3.1 yields the following result:

Corollary 3.1 *Consider a closed set $C \subset \mathbb{R}^n$ defined as the zero superlevel set of a continuously differentiable function $h : \mathbb{R}^n \to \mathbb{R}$ as in (3.3) and assume that $\nabla h(x) \neq 0$ for all $x \in \partial C$. Then C is forward invariant for (2.1) if and only if, for all $x \in \partial C$,*

$$L_f h(x) \geq 0. \tag{3.5}$$

The above condition can be useful for verifying the safety of a dynamical system as it only requires checking that (3.5) holds on the boundary of C. However, the fact that this condition is defined only on the boundary of C makes it challenging to use as the basis for a control design. Ideally, it would be useful to establish invariance conditions over the entirety of C (or even all of $\mathbb{R}^n$) so that a single continuous controller $u = k(x)$ could be used to enforce safety, and possibly other control objectives. One approach would be to simply ensure that (3.5) holds over all of C; however, this is restrictive as such an approach would render every superlevel set of h forward invariant (rather than only the zero level set). Extending condition (3.5) to the entirety of C in the least restrictive fashion requires the introduction of a new comparison function.

Definition 3.3 (*Extended class $\mathcal{K}$, $\mathcal{K}_\infty$ functions*) A continuous function $\alpha : (-b, a) \to \mathbb{R}$, $a, b \in \mathbb{R}_{>0}$ is said to be an extended class $\mathcal{K}$ function, denoted by $\alpha \in \mathcal{K}^e$, if $\alpha(0) = 0$ and $\alpha(\cdot)$ is strictly increasing. A continuous function $\alpha : \mathbb{R} \to \mathbb{R}$ is said to be an *extended class $\mathcal{K}_\infty$ function*, denoted by $\alpha \in \mathcal{K}_\infty^e$, if $\alpha(0) = 0$, $\alpha(\cdot)$ is strictly increasing, $\lim_{r \to \infty} \alpha(r) = \infty$, and $\lim_{r \to -\infty} \alpha(r) = -\infty$.

Essentially, an extended class $\mathcal{K}_\infty$ function is a class $\mathcal{K}$ function defined on the entire real line, and facilitates the definition of a *barrier function*, which plays a role dual to that of a Lyapunov function for establishing forward invariance and safety.

Definition 3.4 (*Barrier function*) Let $h : \mathbb{R}^n \to \mathbb{R}$ be a continuously differentiable function defining a set $C \subset \mathbb{R}^n$ as in (3.3) such that $\nabla h(x) \neq 0$ for all $x \in \partial C$. Then, h is said to be a *barrier function* for (2.1) on C if there exists an $\alpha \in \mathcal{K}_\infty^e$ such that for all $x \in \mathbb{R}^n$

$$L_f h(x) \geq -\alpha(h(x)). \tag{3.6}$$

---

[2] The condition that $\nabla h(x) \neq 0$ for all $x \in \partial C$ is equivalent to 0 being a regular value of h, which ensures that $h^{-1}(\{0\}) = \partial C$ is an embedded submanifold of $\mathbb{R}^n$.

Introducing an extended class $\mathcal{K}_\infty$ function on the right-hand-side of (3.6) allows $t \mapsto h(x(t))$ to decrease along trajectories $t \mapsto x(t)$ of (2.1) but never become negative. That is, the condition in (3.6) allows the system trajectory to approach the boundary of C, but never cross it, thereby ensuring safety in a minimally restrictive fashion. This extension is subtle, but plays an important role in extending barrier functions to control systems, as it directly enlarges the class of functions that can serve as a barrier function, thereby directly enlarging the set of controllers that can be used to render the system safe.

Remark 3.1 The requirement in Definition 3.4 that the barrier condition (3.6) holds on all of $\mathbb{R}^n$ can be generalized so that the condition is only required to hold on some open set $\mathcal{D} \supset C$. This generalization also permits the use of an extended $\mathcal{K}^e$ function on the right-hand-side of the inequality in (3.6) provided that such a function is defined on all of $\mathcal{D}$.

The following result constitutes the main result with regard to barrier functions and shows that the existence of such a function is sufficient to certify the forward invariance of C.

Theorem 3.2 *If h is a barrier function for (2.1), then C is forward invariant.*

Proof The properties of α and (3.6) ensure that for $x \in \partial C$, $L_f h(x) \geq 0$ and the conclusion of the theorem follows from Theorem 3.1.   □

One may note that in Definition 3.4, condition (3.6) is required to hold over $\mathbb{R}^n$ (or on some open set $\mathcal{D} \supset C$ as noted in Remark 3.1), rather than only on C. The benefit of enforcing such a condition over a larger set containing C is that it endows the barrier function with a certain degree of robustness in the sense that trajectories that begin outside C will asymptotically approach C in the limit as time goes to infinity. To formalize this idea we introduce the notion of stability with respect to sets.

Definition 3.5 A closed forward invariant set $C \subset \mathbb{R}^n$ is said to be stable for (2.1) if for each $\varepsilon \in \mathbb{R}_{>0}$ there exists a $\delta \in \mathbb{R}_{>0}$ such that

$$\|x_0\|_C < \delta \implies \|x(t)\|_C < \varepsilon, \quad \forall t \in \mathbb{R}_{\geq 0}. \tag{3.7}$$

Definition 3.6 A closed forward invariant set $C \subset \mathbb{R}^n$ is said to be asymptotically stable for (2.1) if it is stable and δ is chosen such that

$$\|x_0\|_C < \delta \implies \lim_{t \to \infty} \|x(t)\|_C = 0. \tag{3.8}$$

Using the above definitions of stability with respect to sets allows for establishing the following result.

Proposition 3.1 *Let h be a barrier function for (2.1) on a set $C \subset \mathbb{R}^n$ as in (3.3) and suppose either one of the following conditions hold:*

- *the vector field f is forward complete;*
- *the set C is compact.*

Then, C is asymptotically stable for (2.1).

3.2 Control Barrier Functions

In this section we discuss an extension of barrier functions to control systems of the form (2.10)

$$\dot{x} = f(x) + g(x)u,$$

via the notion of a *control barrier function* (CBF). Before introducing CBFs we must formally define what it means for a control system with inputs (as opposed to a closed-loop system) to be safe on a set C. Note that the definition of forward invariance introduced in the previous section cannot be directly applied to (2.10) as the trajectories of (2.10) cannot be determined without first fixing a controller. Rather than fixing a feedback controller for (2.10) and then studying the safety of the closed-loop system, we wish to study the intrinsic properties of (2.10) and determine if it is possible to design a feedback controller that renders a set forward invariant for the resulting closed-loop system. Such a property is captured using the notion of *controlled invariance*.

Definition 3.7 (*Controlled invariance*) A set $C \subset \mathbb{R}^n$ is said to be *controlled invariant* for (2.10) if there exists a locally Lipschitz feedback controller $k : \mathbb{R}^n \to \mathcal{U}$ such that C is forward invariant for the closed-loop system $\dot{x} = f(x) + g(x)k(x)$. If C is controlled invariant for (2.10) then C is said to be a *safe set*.

Analogous to how barrier functions certify the forward invariance of sets for closed-loop systems, CBFs certify the controlled invariance, and thus the safety, of sets for control systems.

Definition 3.8 (*Control barrier function*) Let $h : \mathbb{R}^n \to \mathbb{R}$ be a continuously differentiable function defining a set $C \subset \mathbb{R}^n$ as in (3.3) such that $\nabla h(x) \neq 0$ for all $x \in \partial C$. Then, h is said to be a *control barrier function* (CBF) for (2.10) on C if there exists an $\alpha \in \mathcal{K}^e_\infty$ such that for all $x \in \mathbb{R}^n$

$$\sup_{u \in \mathcal{U}} \{L_f h(x) + L_g h(x)u\} > -\alpha(h(x)). \tag{3.9}$$

In other words, h is a CBF for (2.10) if for each $x \in \mathbb{R}^n$ there exists an input $u \in \mathcal{U}$ satisfying the barrier condition

$$L_f h(x) + L_g h(x) u \geq -\alpha(h(x)).$$

As with the definition of barrier functions from the previous section, the CBF condition (3.9) can be generalized to hold on some open set $\mathcal{D} \supset C$ rather than all of $\mathbb{R}^n$ (see Remark 3.1).

Similar to CLFs, the definition of a CBF allows for defining the set-valued map K_{cbf} : $\mathbb{R}^n \rightrightarrows \mathcal{U}$ as

$$K_{\mathrm{cbf}}(x) = \{u \in \mathcal{U} \mid L_f h(x) + L_g h(x) u \geq -\alpha(h(x))\}, \tag{3.10}$$

which assigns to each $x \in \mathbb{R}^n$ a set $K_{\mathrm{cbf}}(x) \subset \mathcal{U}$ of control values satisfying condition (3.9). We again note the distinction between the strict inequality used in the definition of a CBF (3.9) and the nonstrict inequality used in the definition of K_{cbf} (3.10). The purpose of the strict inequality is twofold: (1) the strict inequality helps establish Lipschitz continuity of the resulting QP-based controller; (2) the strict inequality ensures that the supremum in (3.9) can actually be achieved by a given controller. Similar to CLFs, determining if h is a CBF depends heavily on the behavior of $L_g h$. When $\mathcal{U} = \mathbb{R}^m$, the condition in (3.9) is equivalent to

$$\forall x \in \mathbb{R}^n : L_g h(x) = 0 \implies L_f h(x) > -\alpha(h(x)), \tag{3.11}$$

implying that whenever $L_g h(x) \neq 0$ one can always pick some u to satisfy the CBF condition from (3.9). Determining the validity of h as a CBF when $\mathcal{U} \neq \mathbb{R}^m$ is a much more challenging problem–some insights are provided in Section 3.4. The main theoretical result with regard to CBFs is that the existence of such a function implies the controlled invariance of C.

Theorem 3.3 *Let $h : \mathbb{R}^n \rightarrow \mathbb{R}$ be a CBF for (2.10) on a set $C \subset \mathbb{R}^n$. Then, any locally Lipschitz controller $u = k(x)$ satisfying $k(x) \in K_{cbf}(x)$ for all $x \in \mathbb{R}^n$ renders C forward invariant for the closed-loop system.*

Proof With $u = k(x)$, the closed-loop vector field $f_{\mathrm{cl}}(x) := f(x) + g(x)k(x)$ is locally Lipschitz and satisfies $L_{f_{\mathrm{cl}}}(x) \geq -\alpha(h(x))$ for all $x \in \mathbb{R}^n$. Hence, h is barrier function for $\dot{x} = f_{\mathrm{cl}}(x)$ and forward invariance follows from Theorem 3.2. $\qquad\qquad\square$

One of the practical benefits of CBFs is their ability to act as a *safety filter* for a predesigned control policy $k_0 : \mathbb{R}^n \rightarrow \mathbb{R}^m$, which may not have been designed to guarantee safety a priori. In particular, CBFs allow one to modify such a policy in a minimally invasive fashion to guarantee safety by filtering out unsafe actions from k_0 through the optimization-based controller

$$
\begin{aligned}
k(x) = \arg\min_{u \in \mathcal{U}} \quad & \frac{1}{2}\|u - k_0(x)\|^2 \\
\text{subject to} \quad & L_f h(x) + L_g h(x) u \geq -\alpha(h(x)),
\end{aligned}
\tag{3.12}
$$

which is a QP for a given x if $\mathcal{U} = \mathbb{R}^m$ or $\mathcal{U}$ is a convex polytope. For the controller in (3.12) to provide safety guarantees, it is important that it is Lipschitz continuous as required by Theorem (3.9). Fortunately, when $\mathcal{U} = \mathbb{R}^m$ the QP in (3.12) is a special case of (2.20), and thus has a closed-form solution given by

$$k(x) = \begin{cases} k_0(x) & \text{if } \psi(x) \geq 0 \\ k_0(x) - \frac{\psi(x)}{\|L_g h(x)^\top\|^2} L_g h(x)^\top & \text{if } \psi(x) < 0, \end{cases} \tag{3.13}$$

where $\psi(x) := L_f h(x) + L_g h(x) k_0(x) + \alpha(h(x))$, and is locally Lipschitz.

Rather than filtering a pre-designed control policy through the QP in (3.12), it is also possible to unify both CBF (safety) and CLF (performance) objectives in a single QP. Since the constraints corresponding to these objectives may be conflicting (in the sense that there may not exist a single control value satisfy both constraints simultaneously), it is necessary to relax one of the objectives by treating the corresponding constraint as a soft constraint. Since safety objectives are typically "hard" in the sense that they cannot be relaxed, it is common to relax the performance objective (represented by a CLF) by replacing the standard CLF condition $\dot{V} \leq -\gamma(V)$ for some $\gamma \in \mathcal{K}$ with the soft constraint $\dot{V} \leq -\gamma(V) + \delta$, where $\delta \in \mathbb{R}$ is a slack variable. Taking such an approach, one can solve the QP

$$k(x), \delta^* = \arg\min_{u \in \mathcal{U}, \delta \in \mathbb{R}} \frac{1}{2} u^\top H(x) u + F(x)^\top u + p\delta^2$$
$$\text{subject to } L_f h(x) + L_g h(x) u \geq -\alpha(h(x)) \tag{3.14}$$
$$L_f V(x) + L_g V(x) u \leq -\gamma(V(x)) + \delta$$

where $H : \mathbb{R}^n \to \mathbb{R}^{m \times m}$ is locally Lipschitz and $H(x)$ is a positive definite matrix for each $x \in \mathbb{R}^n$, $F : \mathbb{R}^n \to \mathbb{R}^m$ is locally Lipschitz, and $p \in \mathbb{R}_{>0}$ is a weight that penalizes the magnitude of the CLF relaxation, to obtain a controller satisfying the CBF conditions and (relaxed) CLF conditions.

We close this section by providing two example of applications where CBFs are well-suited to address the competing objectives of stability and safety.

Example 3.1 (Adaptive cruise control) A problem that has served as motivation for the development of CBFs is the adaptive cruise control (ACC) problem. This problem considers a vehicle on the road tasked with driving at a specified speed, while maintaining a safe distance behind the preceding vehicle on the road. Solving such a problem requires designing a feedback controller that balances the potentially competing objectives of driving at a desired speed (stability) and maintaining an appropriate distance behind the lead vehicle on the road (safety) and has thus serves as a well-motivated scenario for benchmarking CBF-based controllers.

We formalize the ACC problem by introducing a system with state $x = [v\,d] \in \mathbb{R}^2$, where $v \in \mathbb{R}$ is the velocity of the ego (controlled) vehicle and $d \in \mathbb{R}$ is the distance between the ego vehicle and the lead vehicle. The dynamics of the system are given by

$$\underbrace{\begin{bmatrix} \dot{v} \\ \dot{d} \end{bmatrix}}_{\dot{x}} = \underbrace{\begin{bmatrix} -\frac{1}{M}\left(f_0 + f_1 v + f_2 v^2\right) \\ v_l - v \end{bmatrix}}_{f(x)} + \underbrace{\begin{bmatrix} \frac{1}{M} \\ 0 \end{bmatrix}}_{g(x)} u, \qquad (3.15)$$

where $M \in \mathbb{R}_{>0}$ is the mass of the ego vehicle, $f_0, f_1, f_2 \in \mathbb{R}_{>0}$ are aerodynamic drag coefficients, and $v_l \in \mathbb{R}_{>0}$ is the velocity of the lead vehicle. The control input $u \in \mathbb{R}$ of the system is the wheel force of the ego vehicle. The specification "drive at a desired speed" can be encoded as the stabilization of the point $v = v_d$. To design a nominal controller achieving this stabilization objective, we take a feedback linearization approach by defining the output $y = v - v_d$. Following the steps outlined in Sect. 2.3.1, we construct a nominal controller k_0 whose stability can be certified using the CLF $V(y) = \frac{1}{2}y^2$. The specification "maintain a safe distance behind the lead vehicle" can be encoded through the safe set $C \subset \mathbb{R}^2$ defined as the zero superlevel set of

$$h(x) = d - \tau_d v,$$

where $\tau_d \in \mathbb{R}_{>0}$ is the desired time headway.[3] Note that h is indeed a CBF when $\mathcal{U} = \mathbb{R}$ as

$$L_g h(x) = \underbrace{\begin{bmatrix} -\tau_d & 1 \end{bmatrix}}_{\nabla h(x)^\top} \begin{bmatrix} \frac{1}{M} \\ 0 \end{bmatrix} = -\frac{\tau_d}{M},$$

which satisfies $L_g h(x) \neq 0$ for all $x \in \mathbb{R}^2$ and $\nabla h(x) \neq 0$ for all $x \in \partial C$.

Results from simulations of the closed-loop system under the resulting CBF-based controller from (3.12) are provided in Fig. 3.1. Note that the ego vehicle initially accelerates from its initial velocity to its desired velocity (left plot); however, as the distance between

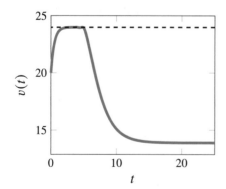
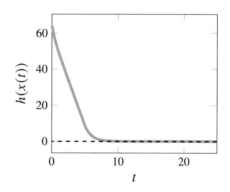

Fig. 3.1 Example simulation of the ACC scenario. The left plot illustrates the evolution of the ego vehicle's velocity with the dashed line indicating the desired velocity $v_d = 24$ m/s. The right plot illustrates the evolution of the CBF along the system trajectory, where the dashed line denotes $h = 0$

---

[3] This is typically chosen as $\tau_d = 1.8$.

the ego and lead vehicle decreases (right plot), the ego vehicle must slow down to ensure it remains a safe distance behind the lead vehicle at all times.

Example 3.2 (**Robot motion planning**) Another scenario in which CBFs have shown success in balancing the competing objectives of performance and safety is in the robot motion planning problem. Here, the objective is to design a feedback controller that drives a robotic system from some initial configuration to a goal configuration while avoiding obstacles in the workspace. To illustrate the application of CBFs to this scenario, we consider an extremely simple version of this problem in which the objective is to drive a planar single integrator $\dot{x} = u$ (i.e., we directly control the robot's velocity) to a goal location while avoiding a circular obstacle in the workspace. The goal-reaching (stability) objective can be accomplished by considering the CLF $V(x) = \frac{1}{2}x^\top x$, whereas the obstacle avoidance (safety) objective can be accomplished by considering the set $C \subset \mathbb{R}^2$ defined as the zero superlevel set of

$$h(x) = \|x - x_o\|^2 - r_o^2,$$

where $x_o \in \mathbb{R}^2$ is the location of the obstacle's center and $r_o \in \mathbb{R}_{>0}$ is its radius. This function is a CBF when $\mathcal{U} = \mathbb{R}^2$ since $\nabla h(x) = 2(x - x_o)$ so that $\nabla h(x) \neq 0$ for $x \in \partial C$, and $L_g h(x) = 2(x - x_o)^\top$, which is non-zero everywhere except for the center of the obstacle. Hence, although the CBF conditions do not hold on all of $\mathbb{R}^n$, they do hold on some open set $\mathcal{D} \supset C$, which is sufficient to guarantee the controlled invariance of C (see Remark 3.1).

We use this simple example to demonstrate the impact of changing the hyperparameters of the CBF, namely, the extended class $\mathcal{K}_\infty$ function α, as well as some limitations of the CBF approach, in general. Common choices of extended class $\mathcal{K}_\infty$ function α include any power function of the form $\alpha(r) = ar^c$, with $a \in \mathbb{R}_{>0}$ and c any odd number, with $c = 1$ or $c = 3$ being favored in practice. In Fig. 3.2 we demonstrate the impact of changing both

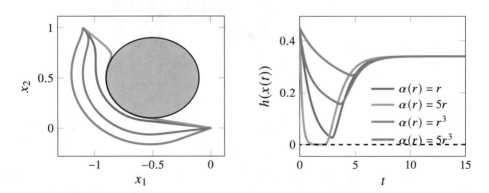

Fig. 3.2 Simulation results for the robot motion planning scenario for different choices of extended class $\mathcal{K}$ function α. The left plot illustrates the trajectories of the closed-loop system, where the gray disk denotes the obstacle, and the right plot shows the evolutions of the CBF along the system dynamics

the coefficient a and power c from α on the resulting trajectory of the closed-loop system under the CBF-QP controller from (3.12), with the nominal policy chosen as the CLF-QP controller. Generally speaking, increasing a leads to more "aggressive" behavior in the sense that the robot is able to approach the obstacle very quickly, whereas smaller values of a result in more conservative behavior. For the particular initial condition used in Fig. 3.2, taking $c = 3$ (i.e., a cubic α) results in more conservative behavior compared to taking $c = 1$ (i.e., a linear α). In general, however, cubic extended class $\mathcal{K}_\infty$ functions allow for quickly approaching the boundary of C when beginning far way, but become more conservative (compared to linear α) near the safe set's boundary.

We close this example with a short discussion on fundamental limitations of the CBF approach in relation to satisfaction of stabilization objectives. When designing a CBF-based controller for the robot motion planning problem, our objective is to assign a vector field to

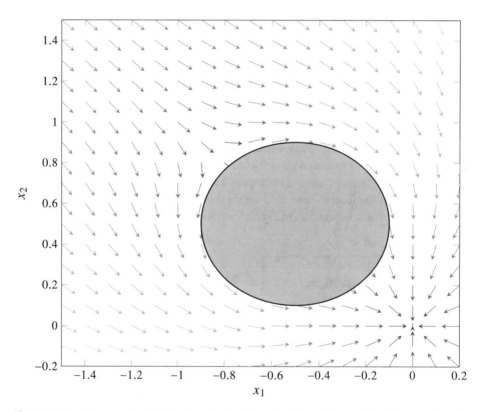

Fig. 3.3 Closed-loop vector field for the robot motion planning problem with $\alpha(r) = r$, where the gray disk denotes the obstacle. The colors of the vectors at each x indicate the magnitude of the closed-loop vector field $f(x) + g(x)k(x)$ evaluated at that point, which is simply the magnitude of the control input at each x. Lighter and darker colors correspond to larger and smaller magnitudes of the control input, respectively

the closed-loop dynamics such that the resulting trajectories flow around an obstacle and to a goal location. This idea is depicted in Fig. 3.3. By inspecting the closed-loop vector field in Fig. 3.3, it appears that simultaneous objectives of stability and safety will be achieved from almost all initial conditions. However, for initial conditions behind the obstacle (i.e., in the top left corner of Fig. 3.3) that satisfy $x_1 = -x_2$, the stabilization objective is not achieved, as the robot gets "stuck" behind the obstacle. We note that this phenomenon is not solely due to the fact that the stabilization objective is treated as "soft" constraint in the QP, but rather due to more fundamental limitations of the behavior that can be exhibited by continuous vector fields. Such limitations have not been extremely detrimental in practice; however, they still must be taken into consideration when designing controllers tasked with guaranteeing both stability and safety.

3.3 High Order Control Barrier Functions

In the previous section we introduced CBFs as a tool to synthesize controllers that guarantee safety, a concept formally encapsulated using the notion of set invariance. Once a CBF is known, it can be used within an optimization-based framework to filter unsafe actions out of an a priori designed controller, or combined with CLFs to generate control actions that mediate the potentially conflicting objectives of performance and safety. Although constructing a controller given a CBF is straightforward, constructing a CBF is a much more challenging problem as it implicitly requires knowledge of a controlled invariant set that can be described as the zero superlevel set of a single continuously differentiable function h.

The challenges in constructing CBFs become more apparent when one makes the distinction between a *state constraint set* and a *safe set*. As discussed in the previous section, a set $C \subset \mathbb{R}^n$ is considered safe for the nonlinear control system (2.10) if it is controlled invariant. That is, a set is a safe set for (2.10) if it is possible to render the set forward invariant for the resulting closed-loop system through the appropriate design of a feedback controller. On the other hand, a state constraint set is simply the set of states that are deemed by the user to not be in violation of a given state constraint. In some simple cases these sets may coincide; in general, however, they may be very different. For example, in the robot motion planning problem introduced in the previous section, the state constraint set corresponds to all positions that are outside the obstacle. This set is also a safe set when the control input is simply the robot's velocity (i.e., a single integrator model) since the state constraint serves as CBF. When the control input is acceleration and the robot's state consists of both position and velocity (i.e., a double integrator model), however, the state constraint set is no longer a controlled invariant set as states on the boundary of such a set will contain velocities that direct the robot into the obstacle.

In this section, we provide a methodology to construct safe sets for nonlinear control systems from user-specified state constraint sets. That is, given a state constraint set $C_0 \subset \mathbb{R}^n$ whose controlled invariance we cannot certify, we aim to produce a controlled invariant set

$C \subset \mathbb{R}^n$ such that $C \subset C_0$. If such a controlled invariant set can be found, then the original state constraint can be conservatively enforced by designing a feedback controller rendering C forward invariant for the closed-loop system. In particular, we consider state constraint sets of the form

$$C_0 = \{x \in \mathbb{R}^n \mid h(x) \geq 0\}, \tag{3.16}$$

where $h : \mathbb{R}^n \rightarrow \mathbb{R}$ is continuously differentiable, which is of the same form as (3.3). As discussed in the previous section, if h is a CBF then clearly such a state constraint set is also a controlled invariant set. However, if $L_g h(x) = 0$ for all $x \in \mathbb{R}^n$, then h is unlikely to be a CBF since the control input has no influence over the behavior of $\dot{h}$. Such a situation arises, for example, when (3.16) represents a constraint on the configuration or kinematics of a mechanical system, but does not take into account the dynamics where the control input will eventually enter. This phenomenon is related to the *relative degree* of h, which we informally introduced in Sect. 2.3 in the context of feedback linearization. We now provide a formal definition of relative degree for a scalar function.

Definition 3.9 (*Relative degree*) A scalar function $h : \mathbb{R}^n \rightarrow \mathbb{R}$ is said to have relative degree $r \in \mathbb{N}$ with respect to (2.10) at a point $x \in \mathbb{R}^n$ if

1. h is r-times continuously differentiable at x;
2. there exists a $\delta \in \mathbb{R}_{>0}$ such that for all $y \in \mathcal{B}_\delta(x)$ and for each $i < r - 1$, we have $L_g L_f^i h(y) = 0$;
3. $L_g L_f^{r-1} h(x) \neq 0$.

If h has relative degree r for each point in a set $\mathcal{D} \subset \mathbb{R}^n$, then h is said to have relative degree r on $\mathcal{D}$.

Remark 3.2 With a slight abuse of terminology, we will often say that a scalar function $h : \mathbb{R}^n \rightarrow \mathbb{R}$ has relative degree $r \in \mathbb{N}$ to mean that it has relative degree r for at least one point in $\mathbb{R}^n$.

We now illustrate many of the ideas introduced in this section thus far using the following running example.

Example 3.3 (**Inverted pendulum**) To illustrate many of the concepts introduced in this section, we now introduce an example that we will revisit later. We consider the inverted pendulum from (2.56) with state $x = [q \ \dot{q}]^\top$, where $q \in \mathbb{R}$ denotes the angular position of the pendulum, and the dynamics are reproduced here for convenience:

$$\dot{x} = \underbrace{\begin{bmatrix} \dot{q} \\ \frac{g}{\ell} \sin(q) - \frac{b}{m\ell^2} \dot{q} \end{bmatrix}}_{f(x)} + \underbrace{\begin{bmatrix} 0 \\ \frac{1}{m\ell^2} \end{bmatrix}}_{g(x)} u.$$

Our objective is to design a feedback controller that ensures the angular position of the pendulum remains less than $\frac{\pi}{4}$ radians. Such an objective can be encapsulated by the constraint

$$h(x) = \tfrac{\pi}{4} - q,$$

which induces a constraint set defined as the zero superlevel set of h. Is it possible to use h as a CBF to render such a constraint set forward invariant? Computing the gradient $\nabla h(x) = [-1\ 0]^\top$ and corresponding Lie derivatives yields

$$L_f h(x) = \begin{bmatrix} -1 & 0 \end{bmatrix} \begin{bmatrix} \dot{q} \\ \frac{g}{\ell} \sin(q) - \frac{b}{m\ell^2}\dot{q} \end{bmatrix} = -\dot{q},$$

$$L_g h(x) = \begin{bmatrix} -1 & 0 \end{bmatrix} \begin{bmatrix} 0 \\ \frac{1}{m\ell^2} \end{bmatrix} = 0.$$

As $L_g h(x) = 0$ for all $x \in \mathbb{R}^n$, the relative degree of h is greater than one for all $x \in \mathbb{R}^n$. That is, it is impossible for the control input to influence the change of h along the system vector field and thus h is not a CBF. The main challenge in using h as a CBF is that, although it defines a relatively simple constraint on the system's configuration, it does not fully capture the behavior necessary to prevent violation of such a constraint. For example, if the pendulum is moving rapidly towards the boundary of the constraint set it may be impossible to stop the pendulum before the constraint is violated. Generally speaking, one must take into account the full dynamics of the system when specifying the safe set via h. For some simple examples, such as the one presented here, there are better choices of h that will yield a valid CBF. In general, however, encoding the behavior necessary to prevent constraint violation in a single continuously differentiable h may be very challenging. In what follows we provide a partial solution to this challenge based upon the observation that, although we may not be able to directly influence the behavior $\dot{h}$, we may be able to control the behavior of higher order Lie derivatives of h. For example, by computing

$$L_g L_f h(x) = \nabla L_f h(x)^\top g(x) = \begin{bmatrix} 0 & -1 \end{bmatrix} \begin{bmatrix} 0 \\ \frac{1}{m\ell^2} \end{bmatrix} = -\frac{1}{m\ell^2},$$

we see that h has relative degree 2 for all $x \in \mathbb{R}^n$, implying that the control input may influence the behavior of $\ddot{h}$ for all $x \in \mathbb{R}^n$.

We provide a solution to the problem outlined in the preceding example by introducing the notion of a *high order control barrier function* (HOCBF). In essence, the HOCBF method works by dynamically extending the original constraint function h and then placing conditions on the higher order derivatives of h that are sufficient to guarantee satisfaction of the original constraint. As will be shown shortly, such an approach provides a natural and systematic method to construct safe sets from user-defined safety constraints.

We begin our exposition by considering a state constraint function $h : \mathbb{R}^n \to \mathbb{R}$ as in (3.16) with relative degree $r \in \mathbb{N}$, which will be dynamically extended to access the control input. That is, we compute the derivative of h along the system dynamics (2.10) until the control input appears. To this end, consider the collection of functions

$$
\begin{aligned}
\psi_0(x) &= h(x), \\
\psi_i(x) &= \dot{\psi}_{i-1}(x) + \alpha_i(\psi_{i-1}(x)), \quad \forall i \in \{1, \ldots, r-1\},
\end{aligned}
\tag{3.17}
$$

where each $\alpha_i \in \mathcal{K}^e_\infty$. Note that as h has relative degree r, each ψ_i for $i \in \{0, \ldots, r-1\}$ is independent of u for all $x \in \mathbb{R}^n$, while $\dot{\psi}_{r-1}(x, u)$ will depend on u at least in some open subset of $\mathbb{R}^n$. We associate to each $\psi_i, i \in \{0, \ldots, r-1\}$, a set $C_i \subset \mathbb{R}^n$ defined as the zero superlevel set of ψ_i:

$$
C_i := \{x \in \mathbb{R}^n \mid \psi_i(x) \geq 0\},
\tag{3.18}
$$

whose intersection

$$
C := \bigcap_{i=0}^{r-1} C_i,
\tag{3.19}
$$

will serve as the set whose controlled invariance we wish to certify using the notion of HOCBF.

Definition 3.10 (*High order CBF*) Let $h : \mathbb{R}^n \to \mathbb{R}$ have relative degree $r \in \mathbb{N}$ for (2.10) that recursively defines a set $C \subset \mathbb{R}^n$ as in (3.19) such that $\nabla\psi_i(x) \neq 0$ for all $x \in \partial C_i$ for each $i \in \{0, \ldots, r-1\}$. Then, h is said to be a *high order control barrier function* (HOCBF) for (2.10) on C if there exists $\alpha_r \in \mathcal{K}^e_\infty$ such that for all $x \in \mathbb{R}^n$

$$
\sup_{u \in \mathcal{U}} \left\{ L_f \psi_{r-1}(x) + L_g \psi_{r-1}(x) u \right\} > -\alpha_r(\psi_{r-1}(x)).
\tag{3.20}
$$

Remark 3.3 The HOCBF condition (3.20) can be expressed in terms of the original constraint function h by noting that

$$
\begin{aligned}
L_f \psi_{r-1}(x) &= L_f^r h(x) + \sum_{i=1}^{r-1} L_f^i (\alpha_{r-i} \circ \psi_{r-i-1})(x), \\
L_g \psi_{r-1}(x) &= L_g L_f^{r-1} h(x).
\end{aligned}
$$

Remark 3.4 As with the CBFs of the previous section, the above definition can be generalized to hold only on some open set $\mathcal{D} \subset \mathbb{R}^n$ containing C rather than all of $\mathbb{R}^n$.

Example 3.4 (**Inverted pendulum (continued)**) Continuing on with our running example, our objective is now to investigate the impact of dynamically extending $h(x) = \frac{\pi}{4} - q$ to compute a candidate safe set as in (3.19). We have already seen that h has relative degree 2

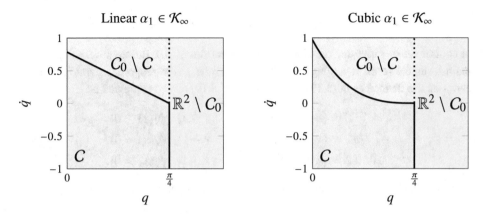

Fig. 3.4 Depiction of the HOCBF induced safe set for the inverted pendulum. The left plot illustrates the resulting safe set when taking α_1 as a linear function whereas the right plot illustrates the resulting safe set when taking α_1 as a cubic function. In each plot, the green region corresponds to the safe set (3.19), the gray region corresponds to states that lie in the state constraint set but not the safe set, and the red region corresponds to states that are in violation of the state constraint. In each plot, the solid black curve denotes the boundary of the safe set. Note that, as the safe set is unbounded, these plots only depict a portion of C–the green region extends infinitely far down the $\dot{q}$ axis and infinitely far to the left along the q axis

for all $x \in \mathbb{R}^n$. We begin by computing the sequence of functions from (3.17) as

$$\psi_0(x) = h(x) = \tfrac{\pi}{4} - q$$
$$\psi_1(x) = \dot{\psi}_0(x) + \alpha_1(\psi_0(x)) = -\dot{q} + \alpha_1\left(\tfrac{\pi}{4} - q\right),$$

for some $\alpha_1 \in \mathcal{K}^e_\infty$. Each of these functions satisfy $\nabla\psi_i(x) \neq 0$ whenever $\psi_i(x) = 0$ and are used to defined two closed sets as

$$C_0 = \{x \in \mathbb{R}^2 \,|\, \tfrac{\pi}{4} - q \geq 0\}$$
$$C_1 = \{x \in \mathbb{R}^2 \,|\, -\dot{q} + \alpha_1\left(\tfrac{\pi}{4} - q\right) \geq 0\}.$$

These sets are used to define a candidate safe set as $C = C_0 \cap C_1$, which is illustrated in Fig. 3.4 for different choices of α_1.

The main idea behind the definition of C is that the dependence of the safety requirement on the higher order dynamics is implicitly encoded through higher order derivatives of h. For this particular example, the resulting safe set C dictates that as the angular position of the pendulum approaches the constraint boundary, the velocity must decrease and eventually become non-positive at the boundary. The choice of extended class $\mathcal{K}_\infty$ function α_1 used in the definition of C determines the manner in which the velocity may decrease.

Following a similar approach to the standard CBF case, a HOCBF allows us to define the set-valued map $K_\psi : \mathbb{R}^n \rightrightarrows \mathcal{U}$ assigning to each $x \in \mathbb{R}^n$ the set

$$K_\psi(x) := \{u \in \mathcal{U} \mid L_f\psi_{r-1}(x) + L_g\psi_{r-1}(x)u \geq -\alpha_r(\psi_{r-1}(x))\}, \tag{3.21}$$

of control values satisfying the HOCBF condition from (3.20). Before proceeding with the main technical result regarding HOCBFs, we require a few properties regarding tangent cones of sets defined as in (3.19). First, note that (3.19) can be expressed as

$$C = \{x \in \mathbb{R}^n \mid \forall i \in \{0, \ldots, r-1\}, \ \psi_i(x) \geq 0\}$$
$$\partial C = \{x \in \mathbb{R}^n \mid \exists i \in \{0, \ldots, r-1\}, \ \psi_i(x) = 0\} \tag{3.22}$$
$$\text{Int}(C) = \{x \in \mathbb{R}^n \mid \forall i \in \{0, \ldots, r-1\}, \ \psi_i(x) > 0\}.$$

Next, denote the set of all *active constraints* of C at a point x by

$$\mathcal{A}_C(x) := \{i \in \{0, \ldots, r-1\} \mid \psi_i(x) = 0\}. \tag{3.23}$$

Then, the tangent cone to C at a point x can be expressed as

$$\mathcal{T}_C(x) = \{v \in \mathbb{R}^n \mid \forall i \in \mathcal{A}_C(x), \ \nabla\psi_i(x)^\top v \geq 0\}, \tag{3.24}$$

provided that $\nabla\psi_i(x) \neq 0$ whenever $\psi_i(x) = 0$ for each $i \in \{0, \ldots, r-1\}$. The following theorem shows that the existence of a HOCBF is sufficient to enforce forward invariance of C and thus satisfaction of the original safety constraint.

Theorem 3.4 *Let $h : \mathbb{R}^n \to \mathbb{R}$ be a HOCBF for (2.10) on a set C as in (3.19). Then, any locally Lipschitz controller $u = k(x)$ satisfying $k(x) \in K_\psi(x)$ for all $x \in \mathbb{R}^n$ renders C forward invariant for the closed-loop system.*

Proof Define $f_{\text{cl}}(x) := f(x) + g(x)k(x)$ as the closed-loop vector field of (2.10) under the controller $u = k(x)$. As the controller is locally Lipschitz, the closed-loop vector field is as well. By the definitions of ψ_i from (3.17) we have

$$\dot{\psi}_{i-1}(x) = L_{f_{\text{cl}}}\psi_{i-1}(x) = \psi_i(x) - \alpha_i(\psi_{i-1}(x)), \quad \forall i \in \{1, \ldots, r-1\},$$

or, equivalently,

$$L_{f_{\text{cl}}}\psi_0(x) = \psi_1(x) - \alpha_1(\psi_0(x))$$
$$L_{f_{\text{cl}}}\psi_1(x) = \psi_2(x) - \alpha_2(\psi_1(x))$$
$$\vdots$$
$$L_{f_{\text{cl}}}\psi_{r-2}(x) = \psi_{r-1}(x) - \alpha_{r-1}(\psi_{r-2}(x)).$$

By (3.22) we have that for all $x \in C$, $\psi_i(x) \geq 0$ for each $i \in \{0, \ldots, r-1\}$. Hence, for all $x \in C$

$$L_{f_{cl}}\psi_0(x) \geq -\alpha_1(\psi_0(x))$$
$$L_{f_{cl}}\psi_1(x) \geq -\alpha_2(\psi_1(x))$$

$$\vdots$$

$$L_{f_{cl}}\psi_{r-2}(x) \geq -\alpha_{r-1}(\psi_{r-2}(x)).$$

By (3.21) the controller $u = k(x) \in K_\psi(x)$ ensures that for all $x \in \mathbb{R}^n$

$$L_{f_{cl}}\psi_{r-1}(x) = L_f\psi_{r-1}(x) + L_g\psi_{r-1}(x)k(x) \geq -\alpha_r(\psi_{r-1}(x)).$$

It thus follows from the two preceding equations that, for all $x \in C$,

$$L_{f_{cl}}\psi_{i-1}(x) \geq -\alpha_i(\psi_{i-1}(x)), \quad \forall i \in \{1, \ldots, r\}.$$

Hence, for all $x \in \partial C$, we have

$$L_{f_{cl}}\psi_i(x) \geq 0, \quad \forall i \in \mathcal{A}_C(x).$$

By the definition of $\mathcal{T}_C$ from (3.24), the preceding argument implies

$$f_{cl}(x) \in \mathcal{T}_C(x), \quad \forall x \in \partial C,$$

and the forward invariance of C follows from Theorem 3.1.   □

Note that the preceding theorem establishes forward invariance of C from (3.19) and not of the constraint set C_0 from (3.16). That is, an initial condition of $x_0 \in C_0$ does not necessarily guarantee that $x(t) \in C_0$ for all $t \in I(x_0)$. Rather, the theorem asserts that $x_0 \in C \implies x(t) \in C$ for all $t \in I(x_0)$, which is sufficient to guarantee that $x(t) \in C_0$ for all $t \in I(x_0)$ since $C \subset C_0$. Similar to CBFs, the motivation behind enforcing the HOCBF conditions on a larger set containing C is to endow the HOCBF with a certain degree of robustness, in the sense that solutions that begin outside or leave[4] C will asymptotically approach C. Indeed, under certain conditions, the set C is also asymptotically stable for the closed-loop system when h is a HOCBF.

Proposition 3.2 *Let h be a HOCBF for (2.10) on a set C as in (3.19) and assume that C is compact. Then, any locally Lipschitz controller $u = k(x)$ satisfying $k(x) \in K_\psi(x)$ for all $x \in \mathbb{R}^n$ renders C asymptotically stable.*

When h is a HOCBF for (2.10), controllers enforcing forward invariance of C can be constructed similarly to the standard CBF case. For example, given a locally Lipschitz nominal control policy $k_0(x)$ one can solve the QP

---

[4] Due to external perturbations/disturbances or unmodeled dynamics.

$$k(x) = \arg\min_{u \in \mathcal{U}} \ \tfrac{1}{2}\|u - k_0(x)\|^2$$
$$\text{subject to } L_f\psi_{r-1}(x) + L_g\psi_{r-1}(x)u \geq -\alpha_r(\psi_{r-1}(x)), \tag{3.25}$$

to produce a controller $u = k(x)$ satisfying $k(x) \in K_\psi(x)$ that acts as a safety filter for the nominal policy. The results regarding Lipschitz continuity of the QP-based controllers in the previous section can be directly extended to show Lipschitz continuity of the HOCBF-QP in (3.25), and to provide a closed form solution to the QP when $\mathcal{U} = \mathbb{R}^m$ as

$$k(x) = \begin{cases} k_0(x) & \text{if } \Psi(x) \geq 0 \\ k_0(x) - \dfrac{\Psi(x)}{\|L_g\psi_{r-1}(x)^\top\|^2} L_g\psi_{r-1}(x)^\top & \text{if } \Psi(x) < 0, \end{cases} \tag{3.26}$$

where $\Psi(x) := L_f\psi_{r-1}(x) + L_g\psi_{r-1}(x)k_0(x) + \alpha_r(\psi_{r-1}(x))$. Of course, the validity of this approach is conditioned upon h being a HOCBF. When $L_g\psi_{r-1}(x) \neq 0$ for all $x \in \mathbb{R}^n$ and $\mathcal{U} = \mathbb{R}^m$, h is a HOCBF since it is always possible to pick an input $u \in \mathbb{R}^m$ satisfying the HOCBF condition (3.20).

Example 3.5 (Inverted pendulum (continued)) We continue our inverted pendulum example by investigating the validity of h as a HOCBF and therefore, by Theorem 3.4, the controlled invariance of C. We have already seen that h has relative degree 2 and that $\nabla\psi_i(x) \neq 0$ for $x \in \partial C_i$ for each $i \in \{0, 1\}$. It thus remains to show that h satisfies the HOCBF condition (3.20), which, by Remark 3.3, can be done by analyzing the behavior of L_gL_fh. As

$$L_gL_fh(x) = -\frac{1}{m\ell^2},$$

which satisfies $L_gL_fh(x) \neq 0$ for all $x \in \mathbb{R}^2$, h is indeed a HOCBF when $\mathcal{U} = \mathbb{R}$. Thus, the controller in (3.25) renders C forward invariant by Theorem 3.4. The closed-loop vector field of the inverted pendulum under such a controller with $\alpha_1(r) = \alpha_2(r) = r$ is provided in Fig. 3.5.

In the preceding example, the validity of h as a HOCBF was guaranteed by ensuring that $L_g\psi_{r-1}h(x) \neq 0$ for all $x \in \mathbb{R}^n$. However, if there exist points in $\mathbb{R}^n$ such that $L_g\psi_{r-1}h(x) = 0$, then (3.20) only holds if

$$\forall x \in \mathbb{R}^n \ : \ L_g\psi_{r-1}(x) = 0 \implies L_f\psi_{r-1}(x) > -\alpha_r(\psi_{r-1}(x)).$$

Unfortunately, the above condition is unlikely to hold at all points in $\mathbb{R}^n$ (or even all points in C) where $L_gL_f^{r-1}h$ vanishes.

Example 3.6 (Singularities in HOCBFs) In this example, we consider the same inverted pendulum as in the previous example, but slightly modify the state constraint to

$$h(x) = \tfrac{\pi}{4} - q^2,$$

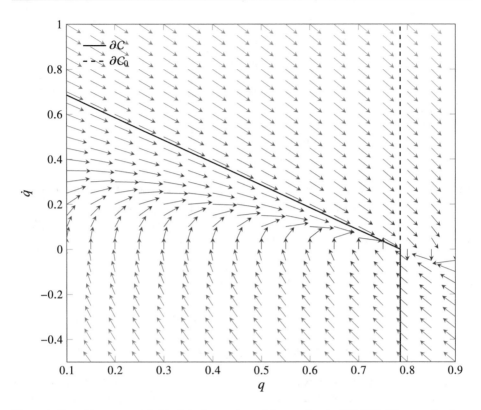

Fig. 3.5 Closed-loop vector field of the inverted pendulum under the HOCBF-QP controller from (3.25) plotted over safe set and constraint set. The dashed black line indicates the boundary of the state constraint set C_0, the solid black line denotes the boundary of the resulting safe set C, and colors of the arrows indicate the magnitude of the vectors

which encodes the requirement that the pendulum's configuration should satisfy $q \in [-\frac{\pi}{4}, \frac{\pi}{4}]$. The gradient of h is

$$\nabla h(x) = \begin{bmatrix} -2q \\ 0 \end{bmatrix},$$

and

$$L_f h(x) = \begin{bmatrix} -2q & 0 \end{bmatrix} \begin{bmatrix} \dot{q} \\ \frac{g}{\ell}\sin(q) - \frac{b}{m\ell^2}\dot{q} \end{bmatrix} = -2q\dot{q}$$

$$L_g h(x) = \begin{bmatrix} -2q & 0 \end{bmatrix} \begin{bmatrix} 0 \\ \frac{1}{m\ell^2} \end{bmatrix} = 0,$$

so h has relative degree larger than one. Computing

$$L_g L_f h(x) = \underbrace{\begin{bmatrix} -2\dot{q} & -2q \end{bmatrix}}_{\nabla L_f h(x)^\top} \begin{bmatrix} 0 \\ \frac{1}{m\ell^2} \end{bmatrix} = -\frac{2q}{m\ell^2},$$

reveals that h has relative degree 2 everywhere except on the set $\{x \in \mathbb{R}^2 \,|\, q = 0\}$. Taking h as a candidate HOCBF we compute

$$\psi_0(x) = \frac{\pi}{4} - q^2$$
$$\psi_1(x) = -2q\dot{q} + \alpha_1 \left(\frac{\pi}{4} - q^2 \right),$$

for some $\alpha_1 \in \mathcal{K}^e_\infty$. The resulting safe set $C = C_0 \cap C_1$ is illustrated in Fig. 3.6 for $\alpha_1(r) = r$, where the dotted red line denotes the set of points where $L_g L_f h$ vanishes.

Since $L_g \psi_1(x) = L_g L_f h(x) = 0$ when $q = 0$, we must ensure that $L_f \psi_1(x) > -\alpha_2(\psi_1(x))$ for some $\alpha_2 \in \mathcal{K}^e_\infty$ at such points. We first note that

$$\nabla \psi_1(x) = \begin{bmatrix} -2\dot{q} - 2q \\ -2q \end{bmatrix},$$

and thus $L_f \psi_1(x) = -2\dot{q}^2$ whenever $q = 0$. Hence, the HOCBF condition requires that

$$q = 0 \implies -2\dot{q}^2 > -\alpha_2(\tfrac{\pi}{4}) \implies \dot{q}^2 < \tfrac{1}{2}\alpha_2(\tfrac{\pi}{4}).$$

That is, the magnitude of velocity must be sufficiently small at points where we cannot directly influence the higher order derivatives of the HOCBF candidate. As such points of high velocity are contained in C, h is not a HOCBF. The inability to satisfy the HOCBF conditions at such points will render the HOCBF-QP (3.25) infeasible at such points, giving rise to singularities in the control input and, consequently, the closed-loop vector field. For example, in the right plot of Fig. 3.6, the HOCBF conditions will be violated along the dotted red line for any points in the shaded regions.

The previous example demonstrates that even relatively simple safety constraints can lead to invalid HOCBFs due to points where $L_g L_f^{r-1} h(x) = 0$. Fortunately, provided the set of points where $L_g L_f^{r-1} h(x) = 0$ does not lie on the boundary of the constraint set ∂C_0, then it is always possible to make a simple modification to h that generates a valid HOCBF.

Proposition 3.3 (Removing the singularity) *Consider a state constraint set $C_0 \subset \mathbb{R}^n$ defined as the zero superlevel set of $h_0 : \mathbb{R}^n \to \mathbb{R}$ as in (3.16) and assume h_0 has relative degree $r \in \mathbb{N}$. Let*

$$\mathcal{E} := \{x \in \mathbb{R}^n \,|\, L_g L_f^{r-1} h_0(x) = 0\},$$

be the set of all points in $\mathbb{R}^n$ where $L_g L_f^{r-1} h_0(x) = 0$ and suppose there exists an $\epsilon \in \mathbb{R}_{>0}$ such that $\mathcal{E}$ is completely contained within the ϵ-superlevel set of h_0:

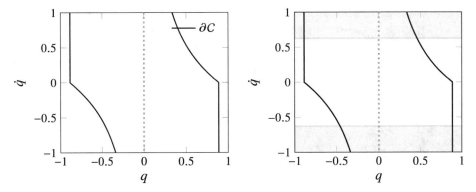

Fig. 3.6 Safe set for the inverted pendulum with a modified safety constraint $h(x) = \frac{\pi}{4} - q^2$. In each plot the safe set is generated using $\alpha_1(r) = r$ whose boundary is depicted by the solid black curve. The dashed red line in each plot indicates the set of points where $L_g L_f h(x) = 0$, which, for this example, is simply $q = 0$. In the right plot the shaded regions represent the set of states where $\dot{q}^2 \geq \frac{1}{2}\alpha_2\left(\frac{\pi}{4}\right)$, which is the set of points where the HOCBF condition is violated at $q = 0$. This set is generated by taking $\alpha_2(r) = r$

$$\mathcal{E} \subset \{x \in \mathbb{R}^n \mid h_0(x) \geq \epsilon\}.$$

Define

$$h(x) := \tau(h_0(x)/\epsilon),\tag{3.27}$$

where $\tau : \mathbb{R} \to \mathbb{R}$ is any sufficiently smooth function satisfying

$$\begin{cases} \tau(0) = 0, \\ \tau(s) = 1, \quad \text{for } s \geq 1 \\ \nabla\tau(s) > 0, \quad \text{for } s < 1. \end{cases}\tag{3.28}$$

Provided $\mathcal{U} = \mathbb{R}^m$, then h from (3.27) is a HOCBF.

The above result exploits the fact that the original invariance conditions introduced in Sect. 3.1 only need to be enforced on the boundary of a given set. The use of extended class $\mathcal{K}$ functions in the CBF and HOCBF approach permits the extension of such conditions to the entirety of a set, but brings additional challenges in verifying that such conditions are satisfied at points where the control input cannot influence the evolution of the CBF/HOCBF. The procedure outlined in Proposition 3.3 essentially modifies the HOCBF candidate h_0 so that the HOCBF conditions trivially hold at points in the constraint set where $L_g L_f^{r-1} h_0(x) = 0$. Provided such points lie strictly in the interior of the constraint set, then no modifications are made to the HOCBF candidate on the boundary of C_0.

Example 3.7 (**Singularities in HOCBFs (continued)** We continue our example of the inverted pendulum with a demonstration of the procedure for generating valid HOCBFs

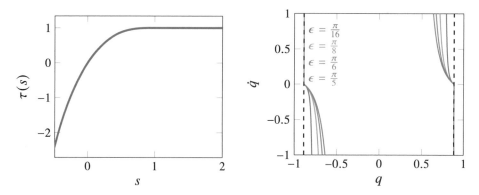

Fig. 3.7 Example of a transition function satisfying the criteria of Proposition 3.3 (left) and modified HOCBF safe set for different values of ϵ (right). Here, the dashed black line denotes the boundary of the state constraint set C_0 and the solid lines of varying color denote the boundary of the resulting safe set C for different values of ϵ as indicated in the legend

outlined in Proposition 3.3. Previously we saw that $h_0(x) = \frac{\pi}{4} - q^2$ is not a valid HOCBF as condition (3.20) fails to hold at all points in the candidate safe set where $L_g L_f h_0(x) = 0$. Such points occur when $q = 0$, which implies that

$$\{x \in \mathbb{R}^2 \mid L_g L_f h_0(x) = 0\} \subset \{x \in \mathbb{R}^2 \mid h_0(x) \geq \epsilon\},$$

for any $\epsilon \in (0, \frac{\pi}{4})$. It follows from Proposition 3.3 that $h(x) = \tau(h_0(x)/\epsilon)$ is a valid HOCBF for any smooth τ satisfying (3.28). An example of such a function when h_0 has relative degree $r = 2$ is given by

$$\tau(s) = \begin{cases} (s-1)^3 + 1, & \text{if } s \leq 1, \\ 1, & \text{if } s > 1, \end{cases}$$

which is illustrated in Fig. 3.7 (left). The safe set resulting from using h as a HOCBF for different values of ϵ is shown in Fig. 3.7 (right).

Although the above approach provides a technique to remove singularities from the HOCBF when the input is unconstrained, the resulting safe set becomes a poor approximation of the viability kernel[5] under input constraints as it allows for extremely high velocities near the boundary of the constraint set (see Fig. 3.7). An alternative, albeit more heuristic, approach is to simply use additional CBFs that prevent the system from entering regions in which the HOCBF conditions are violated as shown in the following example.

Example 3.8 (Singularities in HOCBFs (continued)) We continue our running example of the inverted pendulum by providing a heuristic approach to removing singularities from

---

[5] The viability kernel is the maximum controlled invariant subset of the state constraint set.

Fig. 3.8 The set resulting from intersecting the HOCBF candidate safe set with the candidate safe set that restricts the velocity of the pendulum for different choices of α_2. The curves of varying color denote the boundary of the resulting safe set for different choices of α_2 as noted in the legend

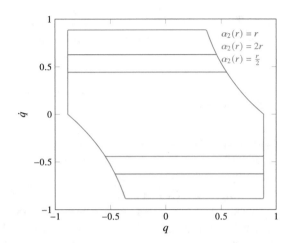

the HOCBF. As we saw previously, the constraint function $h(x) = \frac{\pi}{4} - q^2$ is not a HOCBF for the inverted pendulum (at least not on all of C) because the HOCBF conditions are violated at points in C where $L_g L_f h(x) = 0$. In particular, at such points of singularity, the HOCBF conditions dictate that

$$q = 0 \implies \dot{q}^2 < \tfrac{1}{2}\alpha_2(\tfrac{\pi}{4}),$$

where, recall that for simplicity, we have chosen $\alpha_1(r) = r$ to construct the candidate safe set. Rather than removing the singularity using Proposition 3.3, which results in a safe set that permits extremely large velocities near the boundary of the constraint set, one can alternatively define a new CBF candidate

$$h_v(x) = \tfrac{1}{2}\alpha_2(\tfrac{\pi}{4}) - c - \dot{q}^2,$$

for some arbitrarily small $c \in \mathbb{R}_{>0}$, that defines a set $C_v := \{x \in \mathbb{R}^n \mid h_v(x) \geq 0\}$ whose intersection with the original candidate safe set $C \cap C_v$ represents the subset of C where the HOCBF conditions hold. Note that h_v has relative degree one since

$$\nabla h_v(x) = \begin{bmatrix} 0 \\ -2\dot{q} \end{bmatrix}, \quad L_g h_v(x) = \begin{bmatrix} 0 & -2\dot{q} \end{bmatrix} \begin{bmatrix} 0 \\ \frac{1}{m\ell^2} \end{bmatrix} = -\frac{2\dot{q}}{m\ell^2}.$$

Moreover, h_v is a valid CBF on C_v when $\mathcal{U} = \mathbb{R}^m$ since $\nabla h_v(x) \neq 0$ whenever $h_v(x) = 0$ for c sufficiently small and $L_g h_v(x) = 0$ if and only if $\dot{q} = 0$, which implies that $0 > -\alpha_v(h_v(x))$ whenever $L_g h_v(x) = 0$–a condition that holds everywhere on the interior[6] of C_v. The set resulting from intersecting C with C_v for different choices of α_2 is shown in Fig. 3.8. Although it is possible to individually satisfy the conditions imposed by the

---

[6] Note that a given controller only needs to satisfy the nonstrict inequality to guarantee safety.

HOCBF and h_v within the set depicted in Fig. 3.8, it is not clear if both conditions are *mutually* satisfiable, especially under actuation limits, which is a challenging problem in general.

3.4   Notes

In this chapter, we introduced the notion of safety of dynamical system using the formalism of invariant sets, and how the recent development of control barrier functions (CBFs) facilitates the design of controllers enforcing safety/set invariance. Necessary and sufficient conditions for set invariance were first established by Nagumo in 1942 [1], with similar conditions being independently rediscovered decades later by others [2, 3]. A proof of Nagumo's Theorem (Theorem 3.1) can be found in [4, Ch. 4.1] and a broader introduction to set invariance from a control-theoretic perspective can be found in [5, 6].

The first modern version of a barrier function in the context of safety verification was introduced in [7], where such a function was referred to a *barrier certificate*. Such barrier certificates provided Lyapunov-like sufficient conditions for the verification of invariance properties for nonlinear and hybrid systems. A limitation of these early instantiations of barrier functions is that they effectively enforced the conditions of Corollary 3.1 over the entirety of the safe rather than only on the boundary. Such an approach is overly restrictive as it renders every superlevel set of the safe set forward invariant rather than only the zero superlevel set. Other early attempts to define barrier functions in control theory came via the notion of a *barrier Lyapunov function* (BLF) [8]. These BLFs operate under the same premise as the barrier certificates mentioned above but are also positive definite, which further restricts the class of safe sets that can be described using such a function.

The notion of a barrier function introduced in this chapter first appeared in the series of papers [9–11]. Here, a subtle, yet largely consequential, modification to early definitions of barrier functions was proposed by extending the conditions for set invariance from the boundary of the safe set to the entirety of the safe set in a least-restrictive fashion. In essence, these works replaced the earlier condition

$$\dot{h}(x) \geq 0, \quad \forall x \in C,$$

with the less restrictive condition

$$\dot{h}(x) \geq -\alpha(h(x)), \quad \forall x \in \mathbb{R}^n, \quad \alpha \in \mathcal{K}^e,$$

which allows the value of the barrier function h to decrease along the system trajectory, but never become negative. Originally, such conditions were formulated for *reciprocal* barrier functions [9] that take unbounded values on the boundary of the safe set, but were quickly extended to *zeroing* barrier functions in [10, 11] that vanish on the boundary of the safe set, such as those proposed in Definition 3.4. Shifting from reciprocal to zeroing barrier functions allowed for the development of robustness results for such barriers, such as Proposition 3.1, that ensure the safe set is not only forward invariant, but also asymptotically stable. The concept of a *control barrier function* (CBF) was first introduced in [12]; however, the more recent version presented in this chapter was introduced in [9–11]. These works also made popular the quadratic programming (QP) approach to multi-objective control. The adaptive cruise control example is taken from [9–11]. A further discussion on the history and applications of CBFs can be found in [13].

Initial attempts to extend the CBF methodology to constraints with higher relative degree were first explored in [14, 15] by leveraging a backstepping approach to construct CBFs from relative degree 2 safety constraints. A general approach for constructing high order CBFs (HOCBFs) was first developed in [16] using the notion of an *Exponential Control Barrier Functions* (ECBFs). These ECBFs are essentially the same as the HOCBFs presented earlier in this chapter except that the extended class $\mathcal{K}$ functions are limited to linear functions. The HOCBF approach presented here that allows for the use of general extended class $\mathcal{K}$ functions was first developed in [17, 18]. The observation that the HOCBF conditions may be violated at points in the candidate safe set when the constraint function h does not have a uniform relative degree was first pointed out in [19]. The method of "removing the singularity" from the HOCBF was also proposed in [19] along with additional robustness results that guarantee asymptotic stability of the safe set generated by HOCBFs–Propositions 3.2 and 3.3 were first stated and proved in [19] as well. Alternative methods to constructing CBFs from high relative degree safety constraints rely on extensions of CLF backstepping as introduced in Sect. 2.3.2 to CBF backstepping–a process introduced in [20].

Most methods to construct CBFs do so under the premise that the system's control authority is unlimited (i.e., $\mathcal{U} = \mathbb{R}^m$). Constructing valid CBFs when additional actuation limits are present is a challenging task and an active area of research. For simple systems and constraints, it is often possible to analytically derive a CBF that respects actuation bounds. For example, the authors of [21] construct CBFs for mobile robots modeled as double integrators with acceleration bounds by determining the minimum braking distance required to stop before colliding with obstacles. Such an approach is highly effective in certain scenarios, but generally requires closed-form solutions to the system dynamics. This idea is extended to general systems and constraints in [22–27] using the notion of a *backup* CBF. This approach allows for systematically determining a control invariant subset of the constraint set, which is implicitly represented using a backup control law that renders a smaller set forward invariant. Notably, such an approach is applicable to safety constraints with high relative degree and to systems with actuation bounds, but does require more computational effort at run-time as it requires numerically integrating online the dynamics of the system under the backup policy. Similar ideas in the context of HOCBFs have appeared in [28], whereas [29, 30] provides an alternative approach to constructing HOCBFs under actuation bounds by including additional constraints in the resulting QP that ensure the system remains within a controlled invariant subset of the constraint set. Other approaches to constructing valid CBFs and HOCBFs rely on sum-of-squares programming [31] and machine learning [32, 33].

As briefly noted in Example 3.2, uniting CBFs and CLFs may fail to (asymptotically) stabilize a system when the stability and safety conditions are not mutually satisfiable. In Example 3.2 this manifests itself as the robot getting trapped in a "local minima" behind the obstacle; however, more generally speaking, such a phenomenon arises from the fundamental limitations of the behavior that can be achieved using a continuous static feedback controller. Such limitations were originally pointed out by Brockett in [34], a paper that derived necessary conditions for stabilization by means of continuous static state feedback. Further details on such limitations are discussed in [35, 36]. In the context of CBFs, the work of [37] illustrates that the standard CBF-CLF-QP (3.14) provides no guarantees of stability due to the use of the relaxation variable, even if simultaneous stabilization and safety is possible, and proposes a modification to (3.14) that can be used to establish local asymptotic stability. Additionally, it was shown in [38] the (3.14) may induce additional asymptotically stable points that may be located on the boundary of the safe set. Other works have looked to address these challenges by deriving conditions under which CBFs may be shown to be compatible with CLFs to design controllers that guarantee simultaneous stabilization and safety [39].

References

1. Nagumo N (1942) Uber die lage der integralkurven gewöhnlicher differentialgleichungen. In: Proceedings of the physical-mathematical society of Japan, vol 24, no 3
2. Brezis H (1970) On a characterization of flow-invariant sets. Commun Pure Appl Math 23:261–263
3. Redheffer RM (1972) The theorems of bony and brezis on flow-invariant sets. Am Math Monthly 79(7):740–747
4. Abraham R, Marsden JE, Ratiu T (1988) Manifolds, tensor analysis, and applications, 2nd ed. Springer
5. Blanchini F, Miani S (2008) Set-theoretic methods in control. Springer
6. Blanchini F (1999) Set invariance in control. Automatica 35(11):1747–1767
7. Prajna S, Jadbabaie A (2004) Safety verification of hybrid systems using barrier certificates. In: Proceedings of the international workshop on hybrid systems: computation and control, pp 477–492
8. Tee KP, Ge SS, Tay EH (2009) Barrier lyapunov functions for the control of output-constrained nonlinear systems. Automatica 45(4):918–927
9. Ames AD, Grizzle JW, Tabuada P (2014) Control barrier function based quadratic programs with application to adaptive cruise control. In: Proceedings of the IEEE conference on decision and control, pp 6271–6278
10. Xu X, Tabuada P, Grizzle JW, Ames AD (2015) Robustness of control barrier functions for safety critical control. In: Proceedings of the IFAC conference on analysis and design of hybrid systems, pp 54–61
11. Ames AD, Xu X, Grizzle JW, Tabuada P (2017) Control barrier function based quadratic programs for safety critical systems. IEEE Trans Autom Control 62(8):3861–3876
12. Wieland P, Allgöwer F (2007) Constructive safety using control barrier functions. In: Proceedings of the IFAC symposium on nonlinear control systems

13. Ames AD, Coogan S, Egerstedt M, Notomista G, Sreenath K, Tabuada P (2019) Control barrier functions: theory and applications. In: Proceedings of the European control conference, pp 3420–3431

14. Hsu S, Xu X, Ames AD (2015) Control barrier function based quadratic programs with application to bipedal robotic walking. In: Proceedings of the American control conference, pp 4542–4548

15. Nguyen Q, Sreenath K (2015) Safety-critical control for dynamical bipedal walking with precise footstep placement. In: Proceedings of the IFAC conference on analysis and design of hybrid systems, pp 147–154

16. Nguyen Q, Sreenath K (2016) Exponential control barrier functions for enforcing high relative-degree safety-critical constraints. In: Proceedings of the American control conference, pp 322–328

17. Xiao W, Belta C (2019) Control barrier functions for systems with high relative degree. In: Proceedings of the IEEE conference on decision and control, pp 474–479

18. Xiao W, Belta C (2022) High order control barrier functions. IEEE Trans Autom Control 67(7):3655–3662

19. Tan X, Cortez WS, Dimarogonas DV (2022) High-order barrier functions: robustness, safety and performance-critical control. IEEE Trans Autom Control 67(6):3021–3028

20. Taylor AJ, Ong P, Molnar TG, Ames AD (2022) Safe backstepping with control barrier functions. In: Proceedings of the IEEE conference on decision and control, pp 5775–5782

21. Wang L, Ames AD, Egerstedt M (2017) Safety barrier certificates for collisions-free multirobot systems. IEEE Trans Robot 33(3):661–674

22. Gurriet T, Singletary A, Reher J, Ciarletta L, Feron E, Ames AD (2018) Towards a framework for realizable safety critical control through active set invariance. In: Proceedings of the ACM/IEEE international conference on cyber-physical systems, pp 98–106

23. Guirriet T, Mote M, Ames AD, Feron E (2018) An online approach to active set invariance. In: Proceedings of the IEEE conference on decision and control, pp 3592–3599

24. Gurriet T, Nilson P, Singletary A, Ames AD (2019) Realizable set invariance conditions for cyber-physical systems. In: Proceedings of the American control conference, pp 3642–3649

25. Guirriet T, Mote M, Singletary A, Feron E, Ames AD (2019) A scalable controlled set invariance framework with practical safety guarantees. In: Proceedings of the IEEE conference on decision and control, pp 2046–2053

26. Gurriet T, Mote M, Singletary A, Nilsson P, Feron E, Ames AD (2020) A scalable safety critical control framework for nonlinear systems. IEEE Access, vol 8

27. Chen Y, Jankovic M, Santillo M, Ames AD (2021) Backup control barrier functions: formulation and comparative study. In: Proceedings of the IEEE conference on decision and control, pp 6835–6841

28. Breeden J, Panagou D (2021) High relative degree control barrier functions under input constraints. In: Proceedings of the IEEE conference on decision and control, pp 6119–6124

29. Xiao W, Belta C, Cassandras CG (2022) Sufficient conditions for feasibility of optimal control problems using control barrier functions. Automatica, vol 135

30. Xiao W, Belta C, Cassandras CG (2022) Adaptive control barrier functions. IEEE Trans Autom Control 67(5):2267–2281

31. Clark A (2021) Verification and synthesis of control barrier functions. In: Proceedings of the IEEE conference on decision and control, pp 6105—6112

32. Xiao W, Belta C, Cassandras CG (2020) Feasibility-guided learning for constrained optimal control problems. In: Proceedings of the IEEE conference on decision and control, pp 1896–1901

33. Dawson C, Qin Z, Gao S, Fan C (2021) Safe nonlinear control using robust neural lyapunov-barrier functions. In: Proceedings of the 5th annual conference on robot learning
34. Brockett RW (1983) Asymptotic stability and feedback stabilization. In: Millman RS, Brockett RW, Sussmann H (eds) Differential geometric control theory, pp 181–191. Birkhauser
35. Liberzon D (2003) Switching in systems and control. Birkhäuser, Boston, MA
36. Sontag ED (1999) Stability and stabilization: discontinuities and the effect of disturbances. In: Clarke FH, Stern RJ, Sabidussi G (eds) Nonlinear analysis, differential equations, and control. Springer, Dordrecht, pp 551–598
37. Jankovic M (2018) Robust control barrier functions for constrained stabilization of nonlinear systems. Automatica 96:359–367
38. Reis MF, Aguiar AP, Tabuada P (2021) Control barrier function-based quadratic programs introduce undesirable asymptotically stable equilibria. IEEE Control Syst Lett 5(2):731–736
39. Cortez WS, Dimarogonas DV (2022) On compatibility and region of attraction for safe, stabilizing control laws. IEEE Trans Autom Control 67(9):4924–4931

Adaptive Control Lyapunov Functions

In Chap. 2, we introduced Lyapunov theory as a tool to design controllers enforcing stability of nonlinear control systems. In this chapter, we extend these ideas to nonlinear control systems with *uncertain* dynamics. We focus on the case where the uncertainty enters the system in a structured manner through uncertain parameters. Such a situation occurs when the structure of the vector fields are known, but certain physical attributes of the system (e.g., inertia and damping properties) under consideration are unknown. The dynamics of many relevant systems, especially those in robotics, obey such an assumption. We argue that exploiting this structure, rather than treating uncertainties as a "black box," allows for the development of efficient learning-based approaches to control with strong guarantees of correctness. We provide an introduction to adaptive nonlinear control in Sect. 4.1, where stabilization does not necessarily require convergence of the unknown parameters to their true values. Next, we present two methods that enforce both stability and parameter convergence in Sect. 4.2 and show how convergence can be guaranteed to be exponential in Sect. 4.3. We conclude with final remarks and suggestions for further readings in Sect. 4.4.

Our development in this chapter revolves around a control affine system with additive *parametric* uncertainty

$$\dot{x} = f(x) + F(x)\theta + g(x)u, \tag{4.1}$$

where f and g are as in (2.10), and $F : \mathbb{R}^n \to \mathbb{R}^{n \times p}$ is a matrix-valued function whose columns correspond to vector fields capturing the directions along which the uncertain parameters $\theta \in \mathbb{R}^p$ act.

We assume that $F(0) = 0$ so that the origin remains an equilibrium point of (4.1) with $u = 0$. When the uncertainty enters a system as a linear combination of known nonlinear features $F(x)$ and unknown parameters θ, such as in (4.1), we say that the system is linear in the uncertain parameters or simply linear in the parameters. The model in (4.1) captures

© The Author(s), under exclusive license to Springer Nature Switzerland AG 2023   57
M. Cohen and C. Belta, *Adaptive and Learning-Based Control of Safety-Critical Systems*,
Synthesis Lectures on Computer Science,
https://doi.org/10.1007/978-3-031-29310-8_4

the situation in which we understand the *structure* of the system dynamics, but may not have full knowledge of the system's attributes, such as its inertia or damping properties.

4.1   Adaptive Nonlinear Control

Adaptive control focuses on the problem of simultaneous learning and control: we seek to design a controller $u = k(x, \hat{\theta})$ based on an estimate $\hat{\theta} \in \mathbb{R}^p$ of the uncertain parameters θ, while concurrently improving this estimate using information observed along the system trajectory. This improvement in parameter estimation is accomplished using a parameter update law/estimation algorithm, which manifests itself as a dynamical system $\dot{\hat{\theta}} = \tau(x, \hat{\theta}, t)$, where $\tau : \mathbb{R}^n \times \mathbb{R}^p \times \mathbb{R}_{\geq 0} \to \mathbb{R}^p$ is vector field that is locally Lipschitz in $(x, \hat{\theta})$ and piecewise continuous in t. Thus, the adaptive control problem can be understood as the design of a controller and parameter update law

$$u = k(x, \hat{\theta})$$
$$\dot{\hat{\theta}} = \tau(x, \hat{\theta}, t), \tag{4.2}$$

that guarantee the satisfaction of certain properties (stability, safety) of the closed-loop system. In this regard, adaptive control can be seen as a form of *dynamic* feedback control in the sense that the parameters of the controller adjust over time to accomplish the control objective. A central object studied in the remainder of this book is the parameter estimation error

$$\tilde{\theta} = \theta - \hat{\theta}. \tag{4.3}$$

In general, throughout this book, the notation $\hat{(\cdot)}$ stands for the estimate of some quantity $(\cdot)$, whereas $\tilde{(\cdot)}$ denotes the corresponding estimation error $\tilde{(\cdot)} = (\cdot) - \hat{(\cdot)}$. Note that, since θ is constant, the time-derivative of the parameter estimation error is given by $\dot{\tilde{\theta}} = -\dot{\hat{\theta}}$. As we will see shortly, when studying the behavior of the closed-loop system (4.1), it is often necessary to study the properties of the composite dynamical system

$$\begin{bmatrix} \dot{x} \\ \dot{\tilde{\theta}} \end{bmatrix} = \begin{bmatrix} f(x) + F(x)\theta + g(x)k(x, \hat{\theta}) \\ -\tau(x, \hat{\theta}, t) \end{bmatrix} \tag{4.4}$$

composed of the original system dynamics and the parameter estimation error dynamics.

The traditional problem in adaptive control is to design a dynamic controller that stabilizes (4.1) to an equilibrium point or desired reference trajectory, which can be accomplished using the Lyapunov-based methods introduced in Chap. 2. The following definition provides an extension of CLFs to adaptive control systems.

Definition 4.1 (*Adaptive CLF*) A Lyapunov function candidate V is said to be an *adaptive control Lyapunov function* (aCLF) for (4.1) if there exists $\alpha \in \mathcal{K}$ such that for all $x \in \mathbb{R}^n \setminus \{0\}$ and all $\hat{\theta} \in \mathbb{R}^p$

$$\inf_{u \in \mathcal{U}} \{L_f V(x) + L_F V(x)\hat{\theta} + L_g V(x)u\} < -\alpha(\|x\|). \tag{4.5}$$

As was the case for the CLFs from Chap. 2, an aCLF induces a set-valued map K_{aclf} : $\mathbb{R}^n \times \mathbb{R}^p \rightrightarrows \mathcal{U}$ that associates to each pair $(x, \hat{\theta})$ a set of control values $K_{\text{aclf}}(x, \hat{\theta}) \subset \mathcal{U}$ satisfying the aCLF condition as

$$K_{\text{aclf}}(x, \hat{\theta}) := \{u \in \mathcal{U} \mid L_f V(x) + L_F V(x)\hat{\theta} + L_g V(x)u \leq -\alpha(\|x\|)\}. \tag{4.6}$$

The existence of an aCLF implies the existence of a control policy that stabilizes the estimated dynamics $f(x) + F(x)\hat{\theta} + g(x)u$ for each $\hat{\theta}$. Before observing how such a policy affects the validity of the aCLF V as a Lyapunov function for the actual system dynamics (4.1), we use the parameter estimation error (4.3) to represent (4.1) as

$$\dot{x} = f(x) + F(x)\hat{\theta} + g(x)u + F(x)\tilde{\theta}.$$

This implies the Lie derivative of the aCLF V along the system dynamics can be expressed as

$$\dot{V} = L_f V(x) + L_F V(x)\hat{\theta} + L_g V(x)u + L_F V(x)\tilde{\theta}.$$

Choosing the aCLF induced control policy $u = k(x, \hat{\theta}) \in K_{\text{aclf}}(x, \hat{\theta})$ implies that

$$\dot{V} \leq -\alpha(\|x\|) + L_F V(x)\tilde{\theta}.$$

Unfortunately, the above analysis is inconclusive in terms of stability since the sign-indefinite term $L_F V(x)\tilde{\theta}$ prevents us from drawing any conclusions about the sign of $\dot{V}$. Fortunately, we have yet to specify the other component of our adaptive controller, the parameter update law. In a similar vein to the CLF approach, where we designed a controller to enforce the Lyapunov conditions by construction, in the adaptive control setting we select a parameter update law so that the Lyapunov conditions for the composite system (4.4) are satisfied by construction. To this end, consider a new Lyapunov function candidate for the composite dynamical system (4.4)

$$V_a(x, \tilde{\theta}) = V(x) + \frac{1}{2}\tilde{\theta}^\top \Gamma^{-1}\tilde{\theta}, \tag{4.7}$$

where $\Gamma \in \mathbb{R}^{p \times p}$ is positive definite, which consists of the aCLF V and a weighted quadratic term that penalizes the parameter estimation error. Computing the Lie derivative of V_a along the composite system dynamics with $u = k(x, \hat{\theta}) \in K_{\text{aclf}}(x, \hat{\theta})$ yields

$$\begin{aligned} \dot{V} &= L_f V(x) + L_F V(x)\hat{\theta} + L_g V(x)k(x, \hat{\theta}) + L_F V(x)\tilde{\theta} - \tilde{\theta}^\top \Gamma^{-1}\dot{\hat{\theta}} \\ &\leq -\alpha(\|x\|) + L_F V(x)\tilde{\theta} - \tilde{\theta}^\top \Gamma^{-1}\dot{\hat{\theta}}. \end{aligned} \tag{4.8}$$

Although the problematic term $L_F V(x)\tilde{\theta}$ is still present, we now have another variable $\dot{\hat{\theta}}$ that we are free to choose as we wish. Selecting

$$\dot{\hat{\theta}} = \Gamma L_F V(x)^\top, \tag{4.9}$$

reduces (4.8) to

$$\dot{V} \le -\alpha(\|x\|) \le 0. \tag{4.10}$$

Although we have eliminated all sign-indefinite terms, we can only conclude that $\dot{V} \le 0$ ($\dot{V}$ is negative semidefinite), which implies that the origin of the composite system (4.4) is stable by Theorem 2.2. We can actually go a step further than this and demonstrate convergence of the original system trajectory $t \mapsto x(t)$ to the origin with the help of the following theorem.

Theorem 4.1 (LaSalle-Yoshizawa Theorem) *Consider a dynamical system $\dot{x} = f(x,t)$ where $f : \mathbb{R}^n \times \mathbb{R}_{\ge 0} \to \mathbb{R}^n$ is locally Lipschitz in x, uniformly in t, satisfying $f(0,t) = 0$ for all $t \in \mathbb{R}_{\ge 0}$. Let $V : \mathbb{R}^n \to \mathbb{R}_{\ge 0}$ be a Lyapunov function candidate and assume there exists a continuous function $W : \mathbb{R}^n \to \mathbb{R}$ such that for all $(x,t) \in \mathbb{R}^n \times \mathbb{R}_{\ge 0}$*

$$\dot{V} = L_f V(x,t) \le -W(x) \le 0.$$

Then, all trajectories of $\dot{x} = f(x,t)$ are uniformly bounded and satisfy

$$\lim_{t \to \infty} W(x(t)) = 0.$$

The above result is often referred to as the *LaSalle-Yoshizawa Theorem*. Combining Theorem 4.1 with (4.10) implies that $\lim_{t \to \infty} \alpha(\|x(t)\|) = 0$, which, by the properties of class $\mathcal{K}$ functions, implies $\lim_{t \to \infty} \|x(t)\| = 0$. The preceding discussion is formalized by the following theorem.

Theorem 4.2 *Let V be an aCLF for (4.1) and assume the parameter estimates are updated according to*

$$\dot{\hat{\theta}} = \Gamma L_F V(x)^\top.$$

Then, any controller $u = k(x, \hat{\theta})$ locally Lipschitz on $(\mathbb{R}^n \setminus \{0\}) \times \mathbb{R}^p$ satisfying $k(x, \hat{\theta}) \in K_{aclf}(x, \hat{\theta})$ for all $(x, \hat{\theta}) \in \mathbb{R}^n \times \mathbb{R}^p$ renders the origin of the composite system (4.4) stable and ensures that $\lim_{t \to \infty} \|x(t)\| = 0$.

An interesting consequence of Theorem 4.2 is that the adaptive control strategy is capable of accomplishing the control objective despite there being no guarantee that the estimated parameters converge to their true values. This phenomenon is one of the defining features of adaptive control: exactly learning the parameters is generally not a necessary condition for satisfaction of the control objective.[1]

---

[1] Although, as discussed later in this chapter, there are many benefits to parameter convergence.

In Chap. 2 we saw that for many classes of systems there exist systematic methods to construct CLFs. When system (4.1) satisfies a certain structural condition such techniques can be directly extended to construct aCLFs.

Definition 4.2 (*Matched uncertainty*) The parameters in (4.1) are said to be *matched* if there exists a locally Lipschitz mapping $\varphi : \mathbb{R}^n \to \mathbb{R}^{m \times p}$ such that

$$F(x) = g(x)\varphi(x). \tag{4.11}$$

When the parameters in (4.1) are matched, the feature mapping F can be expressed as a linear combination of the control directions:

$$\dot{x} = f(x) + g(x)(u + \varphi(x)\theta), \tag{4.12}$$

implying that if θ were known then the term $\varphi(x)\theta$ could simply be canceled by the control input. This structural condition on (4.1) greatly simplifies the construction of an aCLF, as it allows one to directly use a CLF for the nominal dynamics $\dot{x} = f(x) + g(x)u$ as an aCLF for the uncertain dynamics (4.1).

Proposition 4.1 *Let V be a CLF for $\dot{x} = f(x) + g(x)u$ with $\mathcal{U} = \mathbb{R}^m$. If the parameters in (4.1) are matched, then V is an aCLF for (4.1) with $\mathcal{U} = \mathbb{R}^m$.*

Proof When $\mathcal{U} = \mathbb{R}^m$, the aCLF condition (4.5) is equivalent to

$$\forall (x, \hat{\theta}) \in \mathbb{R}^n \times \mathbb{R}^p : L_g V(x) = 0 \implies L_f V(x) + L_F V(x)\hat{\theta} < -\alpha(\|x\|). \tag{4.13}$$

If the uncertain parameters are matched, then $F(x) = g(x)\varphi(x)$, which implies that $L_F V(x) = L_g V(x)\varphi(x)$ and, consequently, that $L_g V(x) = 0 \implies L_F V(x) = 0$. Hence, (4.13) reduces to

$$\forall x \in \mathbb{R}^n : L_g V(x) = 0 \implies L_f V(x) < -\alpha(\|x\|), \tag{4.14}$$

which is exactly the statement that V is a CLF for $\dot{x} = f(x) + g(x)u$ when $\mathcal{U} = \mathbb{R}^m$. $\square$

The significance of dealing with matched uncertainty is that the construction of an aCLF can be done independently of the uncertain parameters. When the parameters are not matched the construction of an aCLF is much more challenging, and typically relies on using *adaptive backstepping*, a technique that extends the standard backstepping idea from Chap. 2.3.2 to adaptive control. Note that although we have framed most of the results regarding aCLFs as applicable to the general system (4.1), these are often only applicable to those systems with matched uncertainties. This is because, in general, an aCLF V may also depend on estimates of the uncertain parameters $V(x, \hat{\theta})$, which complicates the aCLF condition (4.5) since, in this situation, the Lie derivative of V depends on the update law $\dot{\hat{\theta}}$, which in turn depends on the aCLF V itself. Since our working definition of an aCLF from Definition 4.1 does not include parameter dependence, most of our discussion in this chapter, and the following few chapters, will be limited to systems with matched parameters. Additional methods to construct aCLFs for systems with unmatched parameters will be discussed in Sect. 4.5, and an adaptive control method that handles general systems of the form (4.1) based on ideas from reinforcement learning will be presented in Chap. 8.

Similar to classical CLFs, once an aCLF is known, a controller satisfying the conditions of Definition 4.1 can be constructed using the QP

$$k(x, \hat{\theta}) = \arg \min_{u \in \mathcal{U}} \quad \frac{1}{2}\|u\|^2$$
$$\text{subject to } L_f V(x) + L_F V(x)\hat{\theta} + L_g V(x)u \leq -\alpha(\|x\|). \tag{4.15}$$

Note that we are not limited to using convex optimization-based controllers–any controller satisfying the criteria of Theorem 4.2 can be used to complete the stabilization task; however, taking the optimization-based approach of (4.15) brings with it the benefits discussed in Chap. 2. When $\mathcal{U} = \mathbb{R}^m$, the QP in (4.15) is a special case of the QP from (2.20) and thus admits the closed form solution

$$k(x, \hat{\theta}) = \begin{cases} 0 & \text{if } \psi(x, \hat{\theta}) \leq 0, \\ -\frac{\psi(x, \hat{\theta})}{\|L_g V(x)^\top\|^2} L_g V(x)^\top & \text{if } \psi(x, \hat{\theta}) > 0, \end{cases} \tag{4.16}$$

where $\psi(x, \hat{\theta}) := L_f V(x) + L_F V(x)\hat{\theta} + \alpha(\|x\|)$, and is locally Lipschitz on $(\mathbb{R}^n \setminus \{0\}) \times \mathbb{R}^p$.

4.2   Concurrent Learning Adaptive Control

In the previous section, we demonstrated how adaptive control provides a methodology to control nonlinear systems with uncertain parameters via online parameter estimation. An interesting property of most adaptive controllers is that convergence of the estimated parameters to their true values is generally not guaranteed. In the Lyapunov-based approach introduced in the previous section, the parameters are updated to satisfy the Lyapunov conditions and not necessarily to provide the best estimate of the parameters. Intuition, however, suggests that better control performance may be achieved if the parameter estimates are driven to their true values. Indeed, parameter convergence in adaptive control is highly desirable as it allows for establishing exponential stability and increases robustness to external disturbances. In this section, we discuss two methods to achieve parameter convergence. The first is based on the notion of persistence of excitation (PE) condition (Sect. 4.2.1). The second is an adaptive control technique that can enforce parameter convergence under conditions that are weaker than PE (Sect. 4.2.2).

4.2.1   Parameter Identification

Establishing parameter convergence in adaptive control has traditionally relied on satisfying the *persistence of excitation* (PE) condition. As the name suggests, such a condition requires the system trajectory to be sufficiently "excited" (a more formal definition is given in Definition 4.3), which typically requires injecting some form of exploration noise in the control input that may detract from performance. To formalize the conditions under which parameter convergence may be achieved, we start by constructing a linear regression model for estimating the uncertain parameters θ. We start by noting that, along a given state-control trajectory $t \mapsto (x(t), u(t))$, the uncertain parameters satisfy the relation

$$\dot{x}(t) - f(x(t)) - g(x(t))u(t) = F(x(t))\theta, \quad \forall t \geq 0.$$

To remove the dependence of the above relation[2] on $\dot{x}$, we note that by integrating the above over a finite time interval $[t - \Delta t, t] \subset \mathbb{R}_{\geq 0}$, $\Delta t \in \mathbb{R}_{\geq 0}$, the relation can be equivalently expressed as

$$x(t) - x(t - \Delta t) - \int_{\max\{t-\Delta t, 0\}}^{t} (f(x(s)) + g(x(s))u(s))\, ds = \int_{\max\{t-\Delta t, 0\}}^{t} F(x(s))\, ds\, \theta,$$

for all $t \geq 0$. Defining

$$\begin{aligned}
\mathcal{Y}(t) &:= x(t) - x(t - \Delta t) - \int_{t\max\{t-\Delta t, 0\}}^{t} (f(x(s)) + g(x(s))u(s))\, ds \\
\mathcal{F}(t) &:= \int_{\max\{t-\Delta t, 0\}}^{t} F(x(s))\, ds,
\end{aligned} \tag{4.17}$$

yields the following linear regression equation for θ:

$$\mathcal{Y}(t) = \mathcal{F}(t)\theta. \tag{4.18}$$

The integrals in (4.17) can be computed using measurements of the system state and control input using standard numerical integration routines. Given a parameter estimate $\hat{\theta}$, we can then compute the prediction error

$$e(\hat{\theta}, t) = \mathcal{Y}(t) - \mathcal{F}(t)\hat{\theta}. \tag{4.19}$$

If the ultimate objective of the parameter estimator were to drive $\hat{\theta}$ to θ, then one possible approach would be to update the estimates to minimize the squared prediction error

---

[2] In practical applications, $\dot{x}$ may contain quantities such as acceleration that may not be directly available for measurement. It is also possible to work with an estimate of the state derivative $\hat{\dot{x}}$; however, this could require numerically differentiating state measurements, which could produce a noisy estimate of $\dot{x}$.

$$E(\hat{\theta}, t) = \tfrac{1}{2}\|e(\hat{\theta}, t)\|^2, \tag{4.20}$$

which could be done using gradient descent

$$\dot{\hat{\theta}} = -\Gamma\frac{\partial E}{\partial\hat{\theta}}(\hat{\theta}, t) = \Gamma\mathcal{F}(t)^\top\left(\mathcal{Y}(t) - \mathcal{F}(t)\hat{\theta}\right), \tag{4.21}$$

where $\Gamma \in \mathbb{R}^{p\times p}$ is positive definite. Using Lyapunov-based tools it is straightforward to show that such an approach ensures that the parameter estimates remain bounded.

Lemma 4.1 *Let $t \mapsto \hat{\theta}(t)$ be generated by (4.21). Then, the parameter estimation error $t \mapsto \tilde{\theta}(t)$ remains bounded for all time.*

Proof Consider the Lyapunov function candidate $V(\tilde{\theta}) = \tfrac{1}{2}\tilde{\theta}^\top\Gamma^{-1}\tilde{\theta}$. The Lie derivative of V along the parameter estimation dynamics can be computed as

$$\begin{aligned}
\dot{V} &= -\tilde{\theta}^\top\Gamma^{-1}\dot{\hat{\theta}} \\
&= -\tilde{\theta}^\top\mathcal{F}(t)^\top\left(\mathcal{Y}(t) - \mathcal{F}(t)\hat{\theta}\right) \\
&= -\tilde{\theta}^\top\mathcal{F}(t)^\top\mathcal{F}(t)\tilde{\theta} \\
&\leq 0,
\end{aligned}$$

where the final inequality follows from the fact that $\mathcal{F}(t)^\top\mathcal{F}(t)$ is at least positive semi-definite. As $\dot{V} \leq 0$, the Lyapunov function candidate is non-increasing along trajectories of the parameter estimation error: $V(\tilde{\theta}(t)) \leq V(\tilde{\theta}(0))$ for all $t \geq 0$. Using the bounds on V it follows that for all $t \geq 0$

$$\|\tilde{\theta}(t)\| \leq \sqrt{\frac{\lambda_{\max}(\Gamma^{-1})}{\lambda_{\min}(\Gamma^{-1})}}\|\tilde{\theta}(0)\| = \sqrt{\frac{\lambda_{\max}(\Gamma)}{\lambda_{\min}(\Gamma)}}\|\tilde{\theta}(0)\|, \tag{4.22}$$

which implies that the parameter estimation error is bounded for all time. □

Similar to the Lyapunov-based parameter estimators outlined in the previous section, there is no guarantee that the parameter estimates will converge to their true values. Traditionally, convergence of the parameter estimates can only be ensured when the system trajectories are *persistently excited*.

Definition 4.3 (*Persistence of excitation*) A matrix-valued signal $\mathcal{F} : \mathbb{R}_{\geq 0} \to \mathbb{R}^{n\times p}$ is said to be *persistently excited* if there exist positive constants $T, c \in \mathbb{R}_{>0}$ such that for all $t \in \mathbb{R}_{\geq 0}$

$$\int_t^{t+T}\mathcal{F}(s)^\top\mathcal{F}(s)ds \geq cI_{p\times p}. \tag{4.23}$$

The persistence of excitation (PE) condition implies that over any given finite time interval $[t, t + T] \subset \mathbb{R}_{\geq 0}$, the matrix $\int_t^{t+T} \mathcal{F}(s)^\top \mathcal{F}(s) ds$ is positive definite, and is historically the condition that has been required for convergence of the parameter estimates in adaptive control. We do not formally show this here, and instead direct the reader to Sect. 4.5 at the end of this chapter for more details. Imposing the PE condition for parameter convergence is challenging for multiple reasons, especially for nonlinear systems. The PE condition cannot be verified a priori since it depends on knowledge of the systems trajectories, which are unknown, and is often not possible to check at run-time since it requires reasoning about all possible future behaviors of the system. From a more practical standpoint, achieving PE often requires exciting the system by injecting a probing signal into the control input, which could cause unexpected behaviors that may be especially undesirable in safety-critical systems.

4.2.2   Concurrent Learning

Concurrent learning is an adaptive control technique that allows for enforcing parameter convergence under much weaker conditions than the PE condition. The main idea behind such an approach is to store input-output data that is observed along the system trajectory in a history stack, which is then leveraged in the parameter update law to facilitate parameter convergence. That is, rather than simply using the data observed at the current time t to update the parameter estimates, as in (4.21), we leverage historical data from previous time steps $t_1 < t_2 < t_3 < \cdots < t$ to improve the parameter estimates. The term "Concurrent Learning" comes from the fact that, in such an approach, instantaneous data is used concurrently with historical data to improve parameter convergence. Intuitively, if this historical data is sufficiently rich (i.e., if the recorded input-output data from previous time-steps are sufficiently distinct from one another), then one can show convergence of the parameters to their true values. We formalize this intuition by defining the notion of a history stack.

Definition 4.4 (*History stack*) A collection of tuples of the form $\mathcal{H}(t) = \{(\mathcal{Y}_j(t), \mathcal{F}_j(t))\}_{j=1}^M$ is said to be a *history stack* with $M \in \mathbb{N}$ entries for system (4.1) at time $t \in \mathbb{R}_{\geq 0}$ if each tuple satisfies

$$\mathcal{Y}_j(t) = \mathcal{F}_j(t)\theta.$$

The piecewise continuous mappings $\mathcal{Y}_j : \mathbb{R}_{\geq 0} \times \mathbb{N} \to \mathbb{R}^n$ and $\mathcal{F}_j : \mathbb{R}_{\geq 0} \times \mathbb{N} \to \mathbb{R}^{n \times p}$ associate to each $t \in \mathbb{R}_{\geq 0}$ and each $j \in \{1, \ldots, M\}$ the values of $\mathcal{Y}(t_i)$ and $\mathcal{F}(t_i)$ as defined in (4.17) recorded at some previous point in time $t_i \leq t$.

A history stack may be initially empty, in which case we define $\mathcal{F}_j(0) = 0, \mathcal{Y}_j(0) = 0$ for all $j \in \{1, \ldots, M\}$, or may contain pre-recorded data from an auxiliary dataset collected offline. Input-output tuples of data generated along a state-control trajectory $t \mapsto (x(t), u(t))$ can be stored in $\mathcal{H}$ by filling in entries that were initially equal to zero, or by replacing previous tuples with new tuples. Algorithms for deciding when to store a given tuple $(\mathcal{Y}(t), \mathcal{F}(t))$ in $\mathcal{H}$ will be discussed shortly. The following definition outlines the key condition that a history stack must satisfy to guarantee convergence of the parameter estimates.

Definition 4.5 (*Finite excitation*) A history stack $\mathcal{H}$ is said to satisfy the *finite excitation* (FE) condition if there exists a time $T \in \mathbb{R}_{\geq 0}$ and a positive constant $\underline{\lambda} \in \mathbb{R}_{>0}$ such that

$$\inf_{t \in [T, \infty)} \left\{ \lambda_{\min} \left(\sum_{j=1}^{M} \mathcal{F}_j(t)^\top \mathcal{F}_j(t) \right) \right\} \geq \underline{\lambda} > 0. \tag{4.24}$$

The above definition states that a history stack $\mathcal{H}$ satisfies the finite excitation (FE) condition if there exists some finite time T such that the matrix $\sum_{j=1}^{M} \mathcal{F}_j(t)^\top \mathcal{F}_j(t)$ is positive definite for all $t \geq T$. The following theorem shows that satisfaction of the finite excitation condition is sufficient for parameter convergence.

Theorem 4.3 *Consider system* (4.1) *and let $\mathcal{H}$ be a history stack for* (4.1). *If the estimated parameters are updated according to*

$$\dot{\hat{\theta}} = \Gamma \sum_{j=1}^{M} \mathcal{F}_j(t)^\top [\mathcal{Y}_j(t) - \mathcal{F}_j(t)\hat{\theta}], \tag{4.25}$$

where $\Gamma \in \mathbb{R}^{p \times p}$ is positive definite, and $\mathcal{H}$ satisfies the finite excitation condition, then the parameter estimation error $t \mapsto \tilde{\theta}(t)$ exponentially converges to zero in the sense that, for all $t \in \mathbb{R}_{\geq 0}$,

$$\|\tilde{\theta}(t)\| \leq \sqrt{\frac{\lambda_{\max}(\Gamma)}{\lambda_{\min}(\Gamma)}} \|\tilde{\theta}(0)\| e^{-\underline{\lambda}\lambda_{\max}(\Gamma)(t-T)}. \tag{4.26}$$

Proof Consider the Lyapunov function candidate $V(\tilde{\theta}) = \frac{1}{2}\tilde{\theta}^\top \Gamma^{-1}\tilde{\theta}$, which satisfies

$$\frac{1}{2\lambda_{\max}(\Gamma)}\|\tilde{\theta}\|^2 = \frac{1}{2}\lambda_{\min}(\Gamma^{-1})\|\tilde{\theta}\|^2 \leq V(\tilde{\theta}) \leq \frac{1}{2}\lambda_{\max}(\Gamma^{-1})\|\tilde{\theta}\|^2 = \frac{1}{2\lambda_{\min}(\Gamma)}\|\tilde{\theta}\|^2, \quad (4.27)$$

for all $\tilde{\theta} \in \mathbb{R}^p$. The Lie derivative of V along the parameter estimation dynamics can be computed as

$$\dot{V}(\tilde{\theta}, t) = -\tilde{\theta}^\top \Gamma^{-1}\dot{\hat{\theta}}$$

$$= -\tilde{\theta}^\top \sum_{j=1}^{M} \mathcal{F}_j(t)^\top \left(\mathcal{Y}_j(t) - \mathcal{F}_j(t)\hat{\theta} \right)$$

$$= -\tilde{\theta}^\top \sum_{j=1}^{M} \mathcal{F}_j(t)^\top \mathcal{F}_j(t)\tilde{\theta}.$$

For any $t \geq 0$, the matrix $\sum_{j=1}^{M} \mathcal{F}_j(t)^{\top} \mathcal{F}_j(t)$ is at least positive semi-definite, implying that $\dot{V}(\tilde{\theta}, t) \leq 0$ for all $t \geq 0$ and thus $V(\tilde{\theta}(t), t) \leq V(\tilde{\theta}(0), 0)$ for all $t \geq 0$. Provided $\mathcal{H}$ satisfies the finite excitation condition, then for all $t \geq T$, the matrix $\sum_{j=1}^{M} \mathcal{F}_j(t)^{\top} \mathcal{F}_j(t)$ is positive definite, allowing $\dot{V}$ to be further bounded as

$$\dot{V}(\tilde{\theta}, t) \leq -\underline{\lambda} \|\tilde{\theta}\|^2 \leq -2\underline{\lambda}\lambda_{\min}(\Gamma)V(\tilde{\theta}, t), \quad \forall t \geq T. \tag{4.28}$$

Invoking the comparison lemma to solve the above differential inequality over the interval $[T, \infty)$ then yields

$$V(\tilde{\theta}(t), t) \leq V(\tilde{\theta}(T), T)e^{-2\underline{\lambda}\lambda_{\min}(\Gamma)(t-T)} \leq V(\tilde{\theta}(0), 0)e^{-2\underline{\lambda}\lambda_{\min}(\Gamma)(t-T)}, \tag{4.29}$$

where the second inequality follows from the fact that $V(\tilde{\theta}(t), t) \leq V(\tilde{\theta}(0), 0)$ for all $t \geq 0$. Note that the above bound is also valid for all $t \geq 0$ as

$$V(\tilde{\theta}(0), 0) \leq V(\tilde{\theta}(0), 0)e^{-2\underline{\lambda}\lambda_{\min}(\Gamma)(t-T)}, \quad \forall t \in [0, T].$$

Combining the bounds in (4.27) with those in (4.29) yields (4.26), as desired. □

The main assumption imposed in the previous theorem is that $\mathcal{H}$ satisfies the FE condition, which requires the existence of a time T such that the matrix $\sum_{j=1}^{M} \mathcal{F}_j(t)^{\top} \mathcal{F}_j(t)$ is positive definite for all time thereafter. In general, ensuring that $\sum_{j=1}^{M} \mathcal{F}_j(t)^{\top} \mathcal{F}_j(t)$ becomes positive definite after some finite time is challenging, since, just like the PE condition, this would require reasoning about future behavior of the system trajectory. However, there exist methods for ensuring that data is added to $\mathcal{H}$ in such a way that $\lambda_{\min}(\sum_{j=1}^{M} \mathcal{F}_j(t)^{\top} \mathcal{F}_j(t))$ is non-decreasing in time. Hence, if one can verify that $\sum_{j=1}^{M} \mathcal{F}_j(t)^{\top} \mathcal{F}_j(t)$ is positive definite at a single instant in time (e.g., by periodically checking its minimum eigenvalue), then one can verify satisfaction of the finite excitation condition. An algorithm for accomplishing this objective is outlined in Algorithm 1. For ease of presenting the algorithm, we define

$$\lambda_{\min}(\mathcal{H}(t)) := \lambda_{\min}\left(\sum_{j=1}^{M} \mathcal{F}_j(t)^{\top} \mathcal{F}_j(t)\right),$$

for a history stack $\mathcal{H}$, and

$$\mathcal{H}_j(t) := (\mathcal{Y}_j(t), \mathcal{F}_j(t)),$$

as the tuple of data present in the jth slot of $\mathcal{H}$ at time t. This algorithm, referred to as the *Singular Value Maximizing Algorithm* takes as inputs an initial history stack $\mathcal{H}(0)$ and a tolerance threshold for adding new data $\varepsilon \in \mathbb{R}_{>0}$. At each time instant t, the algorithm checks if the value of $\mathcal{F}(t)$ is sufficiently distinct from the value of $\mathcal{F}_i(t)$, with $i \in \{1, \ldots, M\}$ denoting the index of the slot most recently updated (initialized to the first slot), according to the tolerance threshold ε. If the current value of $\mathcal{F}(t)$ is sufficiently different from the previously recorded value and $\mathcal{H}$ is not yet full (i.e., if $i < M$), then the current tuple $(\mathcal{Y}(t), \mathcal{F}(t))$ is stored in slot $i + 1$ of $\mathcal{H}$. If the current value of $\mathcal{F}(t)$ is sufficiently different from the previously recorded value and $\mathcal{H}$ is full (i.e., if $i = M$), then the algorithm checks if adding the current tuple $(\mathcal{Y}(t), \mathcal{F}(t))$ to $\mathcal{H}$ will increase $\lambda_{\min}(\mathcal{H}(t))$. This process entails replacing the tuple of data present in each slot with the current tuple $(\mathcal{Y}(t), \mathcal{F}(t))$ and checking if the minimum eigenvalue of the history stack with the new data is larger than the original stack without the data added. If replacing the existing data in a particular slot leads to an increase in $\lambda_{\min}(\mathcal{H}(t))$, then this newly formed history stack replaces the old history stack. If replacing the existing data in multiple slots leads to an increase in $\lambda_{\min}(\mathcal{H}(t))$, then the current tuple $(\mathcal{Y}(t), \mathcal{F}(t))$ is stored in the slot whose replacement results in the largest increase of $\lambda_{\min}(\mathcal{H}(t))$. If it is not possible to increase $\lambda_{\min}(\mathcal{H}(t))$ by replacing existing data with new data, then no changes to the history stack are made, thereby ensuring that $\lambda_{\min}(\mathcal{H}(t))$ is non-increasing in time.

Algorithm 1 Singular Value Maximizing Algorithm

Require: History stack $\mathcal{H}(0)$ at $t = 0$ and a tolerance for adding new data $\varepsilon \in \mathbb{R}_{>0}$

  $i \leftarrow 1$ ▷ Set stack index to 1

  if $\frac{\|\mathcal{F}(t) - \mathcal{F}_i(t)\|}{\|\mathcal{F}_i(t)\|} \geq \varepsilon$ **then**   ▷ Check if current data is different enough

    if $i < M$ **then**   ▷ If stack is not full

      $i \leftarrow i + 1$ ▷ Bump index by 1

      $\mathcal{H}_i(t) \leftarrow (\mathcal{Y}(t), \mathcal{F}(t))$ ▷ Record current data in ith slot of stack

    else   ▷ If stack is full

      $\mathcal{H}_{\text{temp}} \leftarrow \mathcal{H}$ ▷ Copy data in current stack

      $\lambda_{\text{old}} \leftarrow \lambda_{\min}(\mathcal{H}_{\text{temp}}(t))$ ▷ Compute minimum eigenvalue of current stack

      $\Lambda \leftarrow \emptyset$

      for $j \in \{1, \ldots, M\}$ **do**   ▷ For each entry in the stack

        $\mathcal{H}_{\text{temp},j}(t) \leftarrow (\mathcal{Y}(t), \mathcal{F}(t))$ ▷ Replace data in jth slot with current data

        $\lambda_j \leftarrow \lambda_{\min}(\mathcal{H}_{\text{temp}}(t))$ ▷ Compute minimum eigenvalue

        $\Lambda \leftarrow \Lambda \cup \{\lambda_j\}$ ▷ Save minimum eigenvalue

        $\mathcal{H}_{\text{temp}} \leftarrow \mathcal{H}$

      end for

      $\lambda_{\text{new}} \leftarrow \max \Lambda$ ▷ Get the largest minimum eigenvalue

      $k \leftarrow \arg \max \Lambda$

      if $\lambda_{\text{new}} > \lambda_{\text{old}}$ **then**

        $\mathcal{H}_k(t) \leftarrow (\mathcal{Y}(t), \mathcal{F}(t))$ ▷ Replace old data with new data

      end if

    end if

  end if

4.3   Exponentially Stabilizing Adaptive CLFs

In the previous section we demonstrated that the convergence properties of traditional parameter estimation routines could be enhanced by leveraging a history stack of recorded input-output data. When such data is sufficiently rich (as characterized by the minimum eigenvalue of a recorded data matrix), such concurrent learning parameter estimators ensure exponential convergence of the parameter estimation error to zero. In the present section, we exploit this exponential convergence to endow adaptive controllers with exponential stability guarantees. Our development begins by specializing the notion of an aCLF from Definition 4.1.

Definition 4.6 (*Exponentially stabilizing adaptive CLF*) A Lyapunov function candidate $V : \mathbb{R}^n \to \mathbb{R}_{\geq 0}$ is said to be an *exponentially stabilizing adaptive control Lyapunov function* (ES-aCLF) for (4.1) if there exist positive constants $c_1, c_2, c_3 \in \mathbb{R}_{>0}$ such that for all $x \in \mathbb{R}^n$

$$c_1 \|x\|^2 \leq V(x) \leq c_2 \|x\|^2, \tag{4.30}$$

and for all $x \in \mathbb{R}^n \setminus \{0\}$ and $\hat{\theta} \in \mathbb{R}^p$

$$\inf_{u \in \mathcal{U}} \{L_f V(x) + L_F V(x)\hat{\theta} + L_g V(x)u\} < -c_3 \|x\|^2. \tag{4.31}$$

Following the same recipe outlined in Chap. 2, given an ES-aCLF we construct the set-valued map

$$K_{\text{es-aclf}}(x, \hat{\theta}) := \{u \in \mathcal{U} \mid L_f V(x) + L_F V(x)\hat{\theta} + L_g V(x)u \leq -c_3 \|x\|^2\}, \tag{4.32}$$

that assigns to each $(x, \hat{\theta})$ a set of control inputs satisfying the conditions from Definition 4.6. The following theorem demonstrates that any controller satisfying the conditions of Definition 4.6 renders the origin of the composite system (4.4) exponentially stable provided the parameter estimates are updated using a history stack that satisfies the FE condition.

Theorem 4.4 *For system* (4.1), *let* $\mathcal{H}$ *be a history stack and* V *be an ES-aCLF. Suppose the estimated parameters are updated according to*

$$\dot{\hat{\theta}} = \Gamma \left(L_F V(x)^\top + \gamma_c \sum_{j=1}^{M} \mathcal{F}_j(t)^\top \left(\mathcal{Y}_j(t) - \mathcal{F}_j(t)\hat{\theta} \right) \right), \tag{4.33}$$

where $\Gamma \in \mathbb{R}^{p \times p}$ *is positive definite and* $\gamma_c \in \mathbb{R}_{>0}$. *If* $\mathcal{H}$ *satisfies the FE condition, then any controller* $u = k(x, \hat{\theta})$ *locally Lipschitz on* $(\mathbb{R}^n \setminus \{0\}) \times \mathbb{R}^p$ *satisfying* $k(x, \hat{\theta}) \in K_{\text{es-aclf}}(x, \hat{\theta})$ *for all* $(x, \hat{\theta}) \in \mathbb{R}^n \times \mathbb{R}^p$ *renders the origin of the composite system* (4.4) *exponentially stable in the sense that*

$$\left\| \begin{bmatrix} x(t) \\ \tilde{\theta}(t) \end{bmatrix} \right\| \le \sqrt{\frac{\eta_2}{\eta_1}} \left\| \begin{bmatrix} x(0) \\ \tilde{\theta}(0) \end{bmatrix} \right\| e^{-\frac{\eta_3}{2\eta_2}(t-T)}, \quad \forall t \ge 0, \tag{4.34}$$

where

$$\eta_1 := \min\{c_1, \tfrac{1}{2\lambda_{\max}(\Gamma)}\}$$
$$\eta_2 := \max\{c_2, \tfrac{1}{2\lambda_{\min}(\Gamma)}\}$$
$$\eta_3 := \min\{c_3, \gamma_c\underline{\lambda}\}.$$

Proof Consider the composite Lyapunov function candidate

$$V_a(x, \tilde{\theta}) := V(x) + \tfrac{1}{2}\tilde{\theta}^\top \Gamma^{-1}\tilde{\theta},$$

which satisfies

$$\eta_1 \left\| \begin{bmatrix} x \\ \tilde{\theta} \end{bmatrix} \right\|^2 \le V_a(x, \tilde{\theta}) \le \eta_2 \left\| \begin{bmatrix} x \\ \tilde{\theta} \end{bmatrix} \right\|^2, \quad \forall (x, \tilde{\theta}) \in \mathbb{R}^n \times \mathbb{R}^p. \tag{4.35}$$

Computing the Lie derivative of V along the composite system dynamics yields

$$\begin{aligned}
\dot{V}_a(x, \tilde{\theta}, t) &= L_f V(x) + L_F V(x)\hat{\theta} + L_g V(x)k(x, \hat{\theta}) + L_F V(x)\tilde{\theta} \\
&\quad - \tilde{\theta}^\top L_F V(x)^\top - \gamma_c \tilde{\theta}^\top \sum_{j=1}^{M} \mathcal{F}_j(t)^\top \left(\mathcal{Y}_j(t) - \mathcal{F}_j(t)\hat{\theta} \right) \\
&= L_f V(x) + L_F V(x)\hat{\theta} + L_g V(x)k(x, \hat{\theta}) - \gamma_c \tilde{\theta}^\top \sum_{j=1}^{M} \mathcal{F}_j(t)^\top \mathcal{F}_j(t)\tilde{\theta} \\
&\le -c_3\|x\|^2 - \gamma_c \tilde{\theta}^\top \sum_{j=1}^{M} \mathcal{F}_j(t)^\top \mathcal{F}_j(t)\tilde{\theta},
\end{aligned} \tag{4.36}$$

where the last inequality follows from the definition of k. For any $t \ge 0$, the matrix $\sum_{j=1}^{M} \mathcal{F}_j(t)^\top \mathcal{F}_j(t)$ is at least positive semi-definite, allowing (4.36) to be further bounded as

$$\dot{V}_a(x, \tilde{\theta}, t) \le -c_3\|x\|^2 \le 0. \tag{4.37}$$

As $\dot{V}_a$ is negative semi-definite for all t, the origin of the composite system is stable and $V_a(x(t), \tilde{\theta}(t), t) \le V_a(x(0), \tilde{\theta}(0), 0)$ for all $t \ge 0$. Moreover, by Theorem 4.1 we have that $\lim_{t \to \infty} c_3\|x(t)\|^2 = 0$, which implies $\lim_{t \to \infty} \|x(t)\| = 0$. Provided $\mathcal{H}$ satisfies the FE condition, then for all $t \ge T$, the matrix $\sum_{j=1}^{M} \mathcal{F}_j(t)^\top \mathcal{F}_j(t)$ is positive definite, allowing (4.36) to be further bounded as

$$\dot{V}_a(x, \tilde{\theta}, t) \le -c_3\|x\|^2 - \gamma_c\underline{\lambda}\|\tilde{\theta}\|^2 \le -\frac{\eta_3}{\eta_2}V_a(x, \tilde{\theta}), \quad \forall t \ge T. \tag{4.38}$$

Invoking the comparison lemma to solve the above differential inequality over the interval $[T, \infty)$ yields

$$V_a(x(t), \tilde{\theta}(t), t) \le V_a(x(T), \tilde{\theta}(T), T)e^{-\frac{\eta_3}{\eta_2}(t-T)} \le V_a(x(0), \tilde{\theta}(0), 0)e^{-\frac{\eta_3}{\eta_2}(t-T)}, \quad (4.39)$$

where the second inequality follows from the observation that $V_a(x(t), \tilde{\theta}(t), t) \le V_a(x(0),$ $\tilde{\theta}(0), 0)$ for all $t \ge 0$. Note that the above bound is also valid for all $t \ge 0$ as

$$V_a(x(0), \tilde{\theta}(0), 0) \le V_a(x(0), \tilde{\theta}(0), 0)e^{-\frac{\eta_3}{\eta_2}(t-T)}, \quad \forall t \in [0, T].$$

Rearranging terms using the bounds on V_a from (4.35) yields the bound in (4.34), as desired.   □

The previous theorem demonstrates that concurrent learning provides a pathway towards guaranteeing both parameter convergence and exponential stability in the context of adaptive control. An interesting consequence of the above theorem is that asymptotic stability of $x = 0$ is guaranteed regardless of the satisfaction of the FE condition. That is, exploiting the recorded data can only aid in the stabilization task–if the FE condition is not satisfied then the overall control objective is still achieved, albeit not in an exponential fashion.

Similar to the aCLFs of Sect. 4.1, when the parameters in (4.1) are matched, the construction of an ES-aCLF can be performed independently of the uncertainty.

Proposition 4.2 *Let* $V : \mathbb{R}^n \to \mathbb{R}_{\ge 0}$ *be an ES-CLF for* $\dot{x} = f(x) + g(x)u$ *with* $\mathcal{U} = \mathbb{R}^m$. *If the parameters in (4.1) are matched, then* V *is an ES-aCLF for (4.1) with* $\mathcal{U} = \mathbb{R}^m$.

Proof The proof follows the same argument as that of Proposition 4.1.   □

Once an ES-aCLF is known, controllers satisfying the conditions of Definition 4.6 can be computed through the QP

$$k(x, \hat{\theta}) = \underset{u \in \mathcal{U}}{\arg\min} \quad \frac{1}{2}\|u\|^2 \tag{4.40}$$
$$\text{subject to } L_f V(x) + L_F V(x)\hat{\theta} + L_g V(x)u \le -c_3\|x\|^2,$$

which admits a closed-form solution when $\mathcal{U} = \mathbb{R}^m$ that is locally Lipschitz on $(\mathbb{R}^n \setminus \{0\}) \times \mathbb{R}^p$ as

$$k(x, \hat{\theta}) = \begin{cases} 0 & \text{if } \psi(x, \hat{\theta}) \le 0, \\ -\frac{\psi(x, \hat{\theta})}{\|L_g V(x)^\top\|^2} L_g V(x)^\top & \text{if } \psi(x, \hat{\theta}) > 0, \end{cases} \tag{4.41}$$

where $\psi(x, \hat{\theta}) := L_f V(x) + L_F V(x)\hat{\theta} + c_3\|x\|^2$.

4.4   Numerical Examples

Example 4.1 (Inverted pendulum (revisited)) Consider again an unstable nonlinear system in the form of the inverted pendulum from (2.56)

$$m\ell^2\ddot{q} - mg\ell\sin(q) = u - b\dot{q},$$

with state $x = [q\ \dot{q}]^\top \in \mathbb{R}^2$, where we assume that the constants b, g are unknown. Defining the uncertain parameter vector as $\theta := [g\ b]^\top \in \mathbb{R}^2$ allows this system to be put into the form of (4.1) as

$$\dot{x} = \underbrace{\begin{bmatrix} \dot{q} \\ 0 \end{bmatrix}}_{f(x)} + \underbrace{\begin{bmatrix} 0 & 0 \\ \frac{1}{\ell}\sin(q) & -\frac{1}{m\ell^2}\dot{q} \end{bmatrix}}_{F(x)} \underbrace{\begin{bmatrix} g \\ b \end{bmatrix}}_{\theta} + \underbrace{\begin{bmatrix} 0 \\ \frac{1}{m\ell^2} \end{bmatrix}}_{g(x)} u. \tag{4.42}$$

Our main objective is to design an aCLF-based controller that stabilizes the origin of the above system. Note that the uncertainty is matched since there exists a $\varphi : \mathbb{R}^n \to \mathbb{R}^{1\times 2}$ satisfying

$$F(x) = \underbrace{\begin{bmatrix} 0 \\ \frac{1}{m\ell^2} \end{bmatrix}}_{g(x)} \underbrace{\begin{bmatrix} m\ell\sin(q) & -\dot{q} \end{bmatrix}}_{\varphi(x)}, \tag{4.43}$$

which implies that an aCLF can be constructed by finding a CLF for the simple double integrator

$$\dot{x} = \underbrace{\begin{bmatrix} \dot{q} \\ 0 \end{bmatrix}}_{f(x)} + \underbrace{\begin{bmatrix} 0 \\ \frac{1}{m\ell^2} \end{bmatrix}}_{g(x)} u, \tag{4.44}$$

Fig. 4.1 Trajectory of the inverted pendulum under the ES-aCLF controller (blue curve) and aCLF controller (orange) curve. Both trajectories start from an initial condition of $x_0 = [\frac{\pi}{6}\ 0]^\top$ and converge to the origin

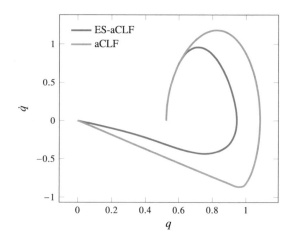

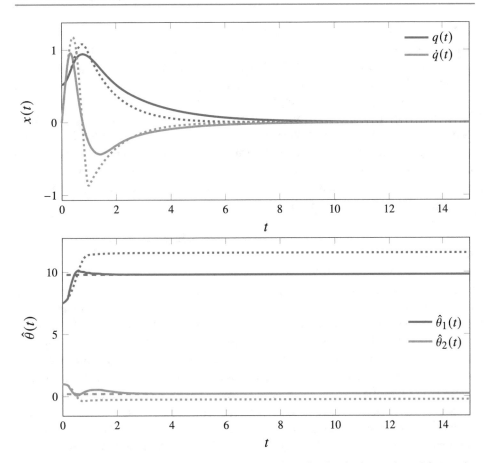

Fig. 4.2 State (top) and parameter estimate (bottom) trajectories for the inverted pendulum under the ES-aCLF controller and aCLF controller. In each plot the solid curves correspond to results generated by the ES-aCLF controller and the dotted curves correspond to those generated by the aCLF controller. In the bottom plot, the dashed lines of corresponding color denote the true values of the unknown parameters

which could be performed using the methods from Sect. 2.3 or by solving a linear quadratic regulator (LQR) problem for the nominal linear system. In what follows we demonstrate numerically the theoretical claim that concurrent learning-based parameter estimators achieve convergence of the parameter estimators under weaker conditions than persistence of excitation. For the simulations we take as our aCLF the function $V(x) = q^2 + \frac{1}{2}\dot{q}^2 + q\dot{q}$ and choose $\Gamma = I_{2 \times 2}$, $\gamma_c = 20$ as learning gains used in the ES-aCLF update law (4.33). The data used in the update law is stored in a history stack with $M = 30$ entries, where the integration window used to generate the data is chosen as $\Delta t = 0.2$.

In the simulations we compare the performance of the ES-aCLF induced controller with that obtained using a standard aCLF (i.e., with no additional data to enforce parameter convergence), the results of which are illustrated in Figs. 4.1 and 4.2. In particular, the plot in Fig. 4.1 depicts the trajectory of the inverted pendulum in the $q \times \dot{q}$ plane, where each trajectory can be seen to converge from its initial condition to the origin. This is also illustrated in Fig. 4.1 (top), which provides the evolution of the pendulum's states over time. The trajectories under each controller are similar; however, the states under the ES-aCLF exhibit less overshoot than those produced by the aCLF controller. Moreover, the parameters under the ES-aCLF controller converge to their true values (see Fig. 4.2 (bottom))–a property that will become very important in the proceeding chapter when extending adaptive control techniques to safety-critical systems.

4.5   Notes

In this chapter we provided a brief introduction to adaptive control of nonlinear systems from a control Lyapunov perspective. Central to our approach was the idea of *concurrent learning* in which instantaneous data is used alongside with a recorded data to learn the uncertain parameters online. Early works in adaptive control primarily focused on the control of uncertain linear systems using the framework of *model reference adaptive control* (MRAC), wherein the objective is to update the parameters of the controller so that the system states converge to those of a suitably constructed reference model. A more in-depth introduction to early results in the adaptive control can be found in several textbooks such as [1–3], with extensions to nonlinear systems developed a few decades later [4–6]. More details on the persistence of excitation (PE) condition can be found in any of the aforementioned textbooks on adaptive control.

Historically, adaptive control methods have been categorized as either *direct* or *indirect*. Direct adaptive control typically parameterizes a controller whose parameters are updated directly to accomplish a control objective. Indirect adaptive control typically estimates the system's uncertain parameters without regard to the underlying control objective, which is then used to compute control actions. The distinction between direct and indirect adaptive control dominates the literature on linear adaptive control; however, for nonlinear systems such a distinction is generally blurred since the parameters used in the controller are typically the estimated system parameters themselves. In the context of nonlinear systems, adaptive control designs are often classified as either *Lyapunov-based*[3] or *estimation-based*. In Lyapunov-based designs, estimates of the uncertain system parameters are typically updated to satisfy the Lyapunov conditions for stability without regard to how accurate such estimates are with respect to the true values of the parameters. On the other hand, estimation-based designs typically update the estimated parameters to minimize the prediction error without considering how such estimates may affect the stability of the closed-loop system.

---

[3] More generally, such designs can be classified as *certificate-based* since they are also applicable to other certificate functions such as barrier functions.

Approaches that combine the benefits of Lyapunov-based and estimation-based (direct and indirect) are often referred to as *composite* adaptive controllers [7].

The control Lyapunov perspective on adaptive control was outlined in [8] with the introduction of the adaptive control Lyapunov function (aCLF). A more recent account of the CLF perspective on adaptive control is presented in [9]. Based on the preceding discussion, the aCLF method is clearly classified as a Lyapunov-based adaptive control approach. Our statement of the LaSalle-Yoshizawa theorem that plays a fundamental role in proving the stability of adaptive control systems is adapted from [6]. A fundamental property that we exploit in this chapter to construct aCLFs is that when the uncertain parameters are matched, an aCLF can be constructed by constructing a CLF for the nominal control affine system. When the parameters are not matched, the construction of an aCLF is more challenging; however, there still exists a wealth of tools for constructing aCLFs in this scenario. The most popular approach is via adaptive backstepping–a process extensively outlined in [6]. Although adaptive backstepping provides a systematic approach towards constructing aCLFs for certain classes of nonlinear systems, the resulting aCLF is often parameter dependent, which significantly complicates the Lyapunov-based update law needed to establish stability. More recently, [10, 11] introduced the notion of an *unmatched* CLF (uCLF), which is a parameter dependent aCLF-type function that can be constructed using the backstepping methodology, but that also allows for the use of much simpler Lyapunov-based update laws. Alternative computational approaches to constructing aCLFs or uCLFs can be performed using sum-of-squares programming provided the system vector fields are polynomial [12].

Traditionally, parameter convergence in adaptive control relied on the PE condition, which, as argued in this chapter, is rather restrictive. Concurrent learning, first introduced by Chowdhary and coauthors [13–15], replaces the PE condition with less restrictive conditions that depend only on data observed along the system trajectory. Exploiting such data allows for ensuring exponential convergence of the parameter estimates to their true values, which, in turn, allows for establishing exponential stability of a composite dynamical system consisting of the system and parameter estimation error dynamics. The singular value maximizing algorithm for recording data that ensures the minimum eigenvalue of the history stack is nonincreasing was first developed in [16]. Based on our earlier discussion, concurrent learning adaptive control can be classified as a composite adaptive control approach. Originally, such concurrent learning techniques required measurements of the state derivative $\dot{x}$ to compute the prediction error–the works of [17, 18] provide a methodology to remove this restriction using state derivative estimation and numerical integration, respectively. Our approach presented in this chapter that alleviates such an assumption by integrating the dynamics over a finite horizon was introduced in [18] and has been referred to as *integral* concurrent learning. The development of concurrent learning adaptive control from a CLF perspective was introduced in [19] using the notion of an exponentially stabilizing aCLF.

References

1. Ioannou PA, Sun J (2012) Robust adaptive control. Dover
2. Ioannou P, Fidan B (2006) Adaptive control tutorial. SIAM
3. Sastry S, Bodson M (2011) Adaptive control: stability, convergence, and robustness. Dover
4. Slotine JJE, Li W (1987) On the adaptive control of robot manipulators. Int J Robot Res 6(3):49–59
5. Slotine JJE, Li W (1991) Applied nonlinear control. Prentice Hall
6. Krstić M, Kanellakopoulos I, Kokotović P (1995) Nonlinear and adaptive control design. Wiley
7. Slotine JJE, Li W (1989) Composite adaptive control of robot manipulators. Automatica 25(4):509–519
8. Krstić M, Kokotović P (1995) Control lyapunov functions for adaptive nonlinear stabilization. Syst Control Lett 26(1):17–23
9. Taylor AJ, Ames AD (2020) Adaptive safety with control barrier functions. In: Proceedings of the American control conference, pp 1399–1405
10. Lopez BT, Slotine JJE (2022) Universal adaptive control of nonlinear systems. IEEE Control Syst Lett 6:1826–1830
11. Lopez BT, Slotine JJE (2022) Adaptive variants of optimal feedback policies. In: 4th annual conference on learning for dynamics and control, vol 166. Proceedings of machine learning research, pp 1–12
12. Moore J, Tedrake, R (2014) Adaptive control design for underactuated systems using sums-of-squares optimization. In: Proceedings of the American control conference, pp 721–728
13. Chowdhary G, Johnson E (2010) Concurrent learning for convergence in adaptive control without persistency of excitation. In: Proceedings of the IEEE conference on decision and control, pp 3674–3679
14. Chowdhary G (2010) Concurrent learning for convergence in adaptive control without persistency of excitation. PhD thesis, Georgia Institute of Technology, Atlanta, GA
15. Chowdhary G, Yucelen T, Muhlegg M, Johnson EN (2013) Concurrent learning adaptive control of linear systems with exponentially convergent bounds. Int J Adapt Control Signal Process 27(4):280–301
16. Chowdhary G, Johnson E (2011) A singular value maximizing data recording algorithm for concurrent learning. In: Proceedings of the American control conference, pp 3547–3552
17. Kamalapurkar R, Reish B, Chowdhary G, Dixon WE (2017) Concurrent learning for parameter estimation using dynamic state-derivative estimators. IEEE Trans Autom Control 62(7):3594–3601
18. Parikh A, Kamalapurkar R, Dixon WE (2019) Integral concurrent learning: adaptive control with parameter convergence using finite excitation. Int J Adapt Control Signal Process 33(12):1775–1787
19. Cohen MH, Belta C (2022) High order robust adaptive control barrier functions and exponentially stabilizing adaptive control lyapunov functions. In: Proceedings of the american control conference, pp 2233–2238

Adaptive Safety-Critical Control

5

In the previous chapter, we discussed how techniques from adaptive control can be used to construct stabilizing controllers for nonlinear systems with uncertain parameters by dynamically adjusting the controller based upon data observed online. In the present chapter, we discuss how the same ideas can be applied to construct controllers that enforce safety, rather than stability specifications. In Sect. 5.1, we define adaptive control barrier functions by extending the notion of adaptive control Lyapunov function that we discussed previously. We further improve this definition by introducing robust adaptive control barrier functions in Sect. 5.2. These notions are first defined for safety constraints with relative degree one with respect to the system dynamics—we extend them for higher relative degrees in Sect. 5.3. We conclude with references, final remarks, and suggestions for further reading in Sect. 5.5.

Throughout this chapter, we focus on the uncertain nonlinear system (4.1) given here again for convenience:

$$\dot{x} = f(x) + F(x)\theta + g(x)u.$$

Our objective is to design an adaptive controller $u = k(x, \hat{\theta})$ that renders a closed set $C \subset \mathbb{R}^n$ forward invariant.

5.1 Adaptive Control Barrier Functions

In this section, we extend the notion of an aCLF from Sect. 4.1 to a safety-critical setting using the notion of an adaptive CBF (aCBF). Once again we consider the set C defined as the zero superlevel set of a continuously differentiable function $h : \mathbb{R}^n \to \mathbb{R}$ as in (3.3):

$$C = \{x \in \mathbb{R}^n \mid h(x) \geq 0\}.$$

As the parameters θ in (4.1) are unknown we cannot directly enforce the CBF condition

© The Author(s), under exclusive license to Springer Nature Switzerland AG 2023   77
M. Cohen and C. Belta, *Adaptive and Learning-Based Control of Safety-Critical Systems*,
Synthesis Lectures on Computer Science,
https://doi.org/10.1007/978-3-031-29310-8_5

$$\sup_{u \in \mathcal{U}} \left\{ L_f h(x) + L_F h(x)\theta + L_g h(x)u \right\} > -\alpha(h(x))$$

over C for some $\alpha \in \mathcal{K}_\infty^e$. Rather, similar to the aCLF approach from Sect. 4.1, we will enforce the CBF condition using an estimated model of the system dynamics and then eliminate the residual parameter estimation by carefully selecting the parameter update law. Such a development motivates the following definition:

Definition 5.1 (*Adaptive CBF*) Let $h : \mathbb{R}^n \to \mathbb{R}$ be a continuously differentiable function defining a set $C \subset \mathbb{R}^n$ as in (3.3) such that $\nabla h(x) \neq 0$ for all $x \in \partial C$. Then, h is said to be an *adaptive control barrier function* (aCBF) for (4.1) if, for all $(x, \hat\theta) \in C \times \mathbb{R}^p$,

$$\sup_{u \in \mathcal{U}} \left\{ L_f h(x) + L_F h(x)\hat\theta + L_g h(x)u \right\} > 0. \tag{5.1}$$

One may notice a few differences between the standard CBF definition and the one presented above, the most notable being the absence of the extended class $\mathcal{K}_\infty$ function on the right-hand-side of the inequality in (5.1). Unfortunately, replacing the right-hand-side of (5.1) with $\alpha(h(x))$ will be insufficient to establish forward invariance of C. Similar to the aCLF case, our analysis will proceed with studying the properties of a *composite* barrier function that contains the parameter estimation error. To this end, consider the composite barrier function candidate

$$h_a(x, \tilde\theta) := h(x) - \tfrac{1}{2}\tilde\theta\Gamma^{-1}\tilde\theta, \tag{5.2}$$

where $\Gamma \in \mathbb{R}^{p \times p}$ is a positive definite learning gain that defines a family of sets $C_\theta \subset \mathbb{R}^n$ parameterized by $\tilde\theta$ as

$$C_\theta = \{x \in \mathbb{R}^n \mid h_a(x, \tilde\theta) \geq 0\}$$
$$\partial C_\theta = \{x \in \mathbb{R}^n \mid h_a(x, \tilde\theta) = 0\} \tag{5.3}$$
$$\text{Int}(C_\theta) = \{x \in \mathbb{R}^n \mid h_a(x, \tilde\theta) > 0\}.$$

Note that $C_\theta \subset C$ for each $\tilde\theta \in \mathbb{R}^p$ as

$$x \in C_\theta \implies h(x) \geq \tfrac{1}{2}\tilde\theta\Gamma^{-1}\tilde\theta \implies h(x) \geq 0 \implies x \in C.$$

Hence, designing a controller that renders C_θ forward invariant provides a pathway towards (conservatively) ensuring that $x(t) \in C$ for all $t \in \mathcal{I}(x_0)$. To facilitate the construction of such a controller, note that an aCBF h induces the set valued map

$$K_{\text{acbf}}(x, \hat\theta) := \{u \in \mathcal{U} \mid L_f h(x) + L_F h(x)\hat\theta + L_g h(x)u \geq 0\}, \tag{5.4}$$

that associates to each $(x, \hat\theta) \in C \times \mathbb{R}^p$ the set of control values $K_{\text{acbf}}(x, \hat\theta) \subset \mathcal{U}$ satisfying the aCBF condition from (5.1). Before stating the main result regarding aCBFs, note that, for a given initial condition $x_0 \in C$ and an initial parameter estimation error $\tilde\theta_0 \in \mathbb{R}^p$, the gain matrix Γ must be selected such that $h(x_0) \geq \tfrac{1}{2}\tilde\theta_0\Gamma^{-1}\tilde\theta_0$ to ensure that $x_0 \in C_\theta$, a sufficient condition for which is that

$$\lambda_{\min}(\Gamma) \geq \frac{\|\tilde{\theta}_0\|^2}{2h(x_0)}. \tag{5.5}$$

The following theorem provides conditions under which a controller drawn from $K_{\mathrm{acbf}}(x, \hat{\theta})$ renders C_θ forward invariant.

Theorem 5.1 *Let $h : \mathbb{R}^n \to \mathbb{R}$ be an aCBF for (4.1) and consider the family of sets $C_\theta \subset C \subset \mathbb{R}^n$ defined by the composite barrier function candidate from (5.2). Suppose the estimated parameters are updated according to*

$$\dot{\hat{\theta}} = -\Gamma L_F h(x)^\top, \tag{5.6}$$

and that Γ is selected such that (5.5) holds for a given initial condition $(x_0, \tilde{\theta}_0) \in C \times \mathbb{R}^p$. Then, the trajectory of the composite dynamical system

$$\begin{bmatrix} \dot{x} \\ \dot{\hat{\theta}} \end{bmatrix} = \begin{bmatrix} f(x) + F(x)\theta + g(x)k(x, \hat{\theta}) \\ \Gamma L_F h(x)^\top \end{bmatrix}, \tag{5.7}$$

with $u = k(x, \hat{\theta}) \in K_{\mathrm{acbf}}(x, \hat{\theta})$ locally Lipschitz on $C \times \mathbb{R}^p$, satisfies $x(t) \in C$ for all $t \in I(x_0, \tilde{\theta}_0)$.

Proof Taking the Lie derivative of the composite barrier function candidate h_a along the vector field of the composite dynamical system (5.7) yields

$$\begin{aligned} \dot{h}_a &= L_f h(x) + L_F h(x)\theta + L_g h(x)k(x, \hat{\theta}) - \tilde{\theta}\Gamma^{-1}\dot{\hat{\theta}} \\ &= L_f h(x) + L_F h(x)\hat{\theta} + L_g h(x)k(x, \hat{\theta}) \geq 0, \end{aligned} \tag{5.8}$$

where the second line follows from substituting in the parameter update law (5.6) and the final inequality follows from the properties of the controller in (5.4). Integrating the above over a finite time interval $[0, t] \subset \mathbb{R}$ reveals that

$$h_a(x(t), \tilde{\theta}(t)) \geq h_a(x_0, \tilde{\theta}_0). \tag{5.9}$$

Hence, provided Γ is selected such that (5.5) holds, then $h_a(x_0, \tilde{\theta}_0) \geq 0$, which implies $x(t) \in C_\theta$ for all $t \in I(x_0, \tilde{\theta}_0)$. As $C_\theta \subset C$ for each $\tilde{\theta} \in \mathbb{R}^p$, the preceding argument implies that $x(t) \in C$ for all $t \in I(x_0, \tilde{\theta}_0)$, as desired. $\qquad \square$

The proceeding theorem provides safety guarantees for the adaptive controller derived from an aCBF by rendering the family of subsets $C_\theta \subset C$ forward invariant. Similar to the definition of a standard CBF, when $\mathcal{U} = \mathbb{R}^m$ the condition in (5.1) can be expressed as

$$\forall(x, \hat{\theta}) \in C \times \mathbb{R}^p : L_g h(x) = 0 \implies L_f h(x) + L_F h(x)\hat{\theta} > 0. \tag{5.10}$$

Verifying the above condition for all $\hat{\theta} \in \mathbb{R}^p$ may be very challenging. Fortunately, when the parameters in (4.1) are matched, the aCBF condition is independent of $\hat{\theta}$.

Proposition 5.1 *Let $h : \mathbb{R}^n \to \mathbb{R}$ be a continuously differentiable function defining a set $C \subset \mathbb{R}^n$ as in (3.3) such that $\nabla h(x) \neq 0$ for all $x \in \partial C$. Suppose $\mathcal{U} = \mathbb{R}^m$ and that (4.1) satisfies the matching condition. Then, h is an aCBF for (4.1) if*

$$\forall x \in C : L_g h(x) = 0 \implies L_f h(x) > 0. \tag{5.11}$$

Proof Follows similar steps to that of Proposition 4.1.   □

Although the above proposition illustrates that the design of h can be decoupled from the uncertain parameters when (4.1) satisfies the matching condition, it does not state that h being a CBF for the nominal dynamics $\dot{x} = f(x) + g(x)u$ is sufficient to guarantee that h is an aCBF for the uncertain dynamics $\dot{x} = f(x) + F(x)\theta + g(x)u$. Indeed, the aCBF condition (5.1) is much stronger than the standard CBF condition (3.9) as it requires *every* non-negative superlevel set of h to controlled invariant rather than only the zero superlevel set, which is one source of conservatism of the aCBF approach. The other source of conservatism stems from the fact that a subset C_θ of C is rendered forward invariant, rather than C itself. As the term $\frac{1}{2}\tilde{\theta}\Gamma^{-1}\tilde{\theta} \to 0$, the composite barrier candidate h_a from (5.2) approaches the original barrier candidate h, which implies $C_\theta \to C$. Hence, the conservatism of the approach can be reduced in two ways: (1) choosing Γ with a larger minimum eigenvalue; (2) decreasing the parameter estimation error $\tilde{\theta}$. In theory, one can take Γ as large as they like; in practice, this is ill-advised as large adaptation gains can amplify the effect of unmodeled dynamics and disturbances. Thus, a more practical approach may be to reduce parameter estimation error, yet, as discussed in the previous chapter, traditional adaptive control methods typically provide no guarantees of convergence of the parameter estimates. Fortunately, the data-driven adaptive control tools introduced in Sect. 4.2 provide a methodology to reduce the level of uncertainty in the parameter estimates online as more data becomes available. As shown in the following section, such tools can be gracefully integrated into the aCBF framework to address the limitations outlined above.

5.2   Robust Adaptive Control Barrier Functions

In this section we demonstrate how the limitations of the aCBF approach can be addressed by uniting tools from concurrent learning adaptive control with the notion of a *robust adaptive control barrier function* (RaCBF). The main idea behind the RaCBF methodology is to robustly account for the worst-case bound on the parameter estimation error, but reduce such a bound online as more data about the system becomes available. For a such an approach to be tractable, we require stronger assumptions on (4.1) in the form of prior knowledge of θ.

Assumption 5.1 There exists a known subset of the parameter space $\Theta \subset \mathbb{R}^p$ and a maximum estimation error $\tilde{\vartheta}_{max} \in \mathbb{R}_{\geq 0}$ such that

$$\Theta := \{\hat{\theta} \in \mathbb{R}^p \mid \|\theta - \hat{\theta}\| \leq \tilde{\vartheta}_{max}\} \tag{5.12}$$

The above assumption implies that, although we do not know the exact values of the uncertain parameters θ, we do know some region of the parameter space in which the parameters lie. From a theoretical standpoint, this is more restrictive than the assumptions posed in the previous chapter in which no assumptions on where the parameters lie were made. We argue, however, that this is not restrictive from a practical standpoint as such parameters generally correspond to the physical attributes of a system (e.g., mass, inertia, damping, etc.) that may take on known ranges of values.

Definition 5.2 (*Robust adaptive CBF*) Let $h : \mathbb{R}^n \to \mathbb{R}$ be a continuously differentiable function defining a set $C \subset \mathbb{R}^n$ as in (3.3) such that $\nabla h(x) \neq 0$ for all $x \in \partial C$. Then, h is said to be a *robust adaptive control barrier function* (RaCBF) for (4.1) on C if there exists $\alpha \in \mathcal{K}_\infty^e$ such that for all $(x, \hat{\theta}) \in \mathbb{R}^n \times \Theta$

$$\sup_{u \in \mathcal{U}} \left\{ L_f h(x) + L_F h(x)\hat{\theta} + L_g h(x)u \right\} > -\alpha(h(x)) + \|L_F h(x)\|\tilde{\vartheta}_{max}. \tag{5.13}$$

When $\mathcal{U} = \mathbb{R}^m$, the RaCBF condition (5.13) can be restated as

$$\forall(x, \hat{\theta}) \in \mathbb{R}^n \times \Theta : L_g h(x) = 0 \implies L_f h(x) + L_F h(x)\hat{\theta}$$
$$> -\alpha(h(x)) + \|L_F h(x)\|\tilde{\vartheta}_{max}, \tag{5.14}$$

which may be very challenging to verify for all possible $\hat{\theta} \in \Theta$. Similar to the results in the previous section, when (4.1) satisfies the matching condition, the criteria for determining the validity of h as a RaCBF becomes much simpler.

Proposition 5.2 *Let h be a CBF for $\dot{x} = f(x) + g(x)u$ with $\mathcal{U} = \mathbb{R}^m$ on a set C and suppose the parameters in (4.1) are matched. Then, h is a RaCBF for (4.1) on C.*

Proof Follows the same steps as that of Proposition 4.1. $\qquad\square$

As in the previous section, an RaCBF induces a family of control policies expressed through the set-valued map

$$K_{RaCBF}(x, \hat{\theta}) := \{u \in \mathcal{U} \mid L_f h(x) + L_F h(x)\hat{\theta} + L_g h(x)u$$
$$\geq -\alpha(h(x)) + \|L_F h(x)\|\tilde{\vartheta}_{max}\}, \tag{5.15}$$

assigning to each $(x, \hat{\theta}) \in \mathcal{D} \times \Theta$ the set $K_{\text{RaCBF}}(x, \hat{\theta}) \subset \mathcal{U}$ of control values satisfying the RaCBF condition (5.13). The following result demonstrates that any locally Lipschitz controller $u = k(x, \hat{\theta})$ satisfying $k(x, \hat{\theta}) \in K_{\text{RaCBF}}(x, \hat{\theta})$ renders C forward invariant.

Proposition 5.3 *Let $h : \mathbb{R}^n \to \mathbb{R}$ be a RaCBF for (4.1) on a set C as in (3.3), and let Assumption 5.1 hold. Then, any locally Lipschitz controller $u = k(x, \hat{\theta})$ satisfying $k(x, \hat{\theta}) \in K_{RaCBF}(x, \hat{\theta})$ for all $(x, \hat{\theta}) \in \mathbb{R}^n \times \Theta$ renders C forward invariant.*

Proof The Lie derivative of h along the closed-loop system vector field is

$$\dot{h} = L_f h(x) + L_F h(x)\theta + L_g h(x) k(x, \hat{\theta})$$
$$= L_f h(x) + L_F h(x)\hat{\theta} + L_g h(x) k(x, \hat{\theta}) + L_F h(x)\tilde{\theta}. \tag{5.16}$$

Provided $\hat{\theta} \in \Theta$ and Assumption 5.1 holds, $\dot{h}$ can be lower bounded as

$$\dot{h} \geq L_f h(x) + L_F h(x)\hat{\theta} + L_g h(x) k(x, \hat{\theta}) - \|L_F h(x)\| \|\tilde{\theta}\|$$
$$\geq L_f h(x) + L_F h(x)\hat{\theta} + L_g h(x) k(x, \hat{\theta}) - \|L_F h(x)\| \tilde{\vartheta}_{\max} \tag{5.17}$$
$$\geq -\alpha(h(x)),$$

where the last inequality follows from (5.15). Thus, since $\nabla h(x) \neq 0$ for all $x \in \partial C$ and $\dot{h} \geq -\alpha(h(x))$, h is a barrier function for the closed-loop system and the forward invariance of C follows from Theorem 3.2. $\qquad \square$

The preceding result shows that a RaCBF remains valid as a safety certificate while updating the parameter estimates by accounting for the worst-case bound on the estimation error. Our ultimate objective, however, is to reduce this worst-case bound over time as our parameter estimates improve. The following result shows that if h is a RaCBF for (4.1) with a given level of model uncertainty $\tilde{\vartheta}_{\max}$, then it remains a RaCBF as the level of uncertainty is reduced.

Lemma 5.1 *If h is a RaCBF for (4.1) for a given $\tilde{\vartheta}_{\max} \in \mathbb{R}_{\geq 0}$, then it is also an RaCBF for any $\tilde{\vartheta} \in [0, \tilde{\vartheta}_{\max}]$ in the sense that for all $(x, \hat{\theta}) \in \mathbb{R}^n \times \Theta$*

$$\sup_{u \in \mathcal{U}} \left\{ L_f h(x) + L_F h(x)\hat{\theta} + L_g h(x)u \right\} > -\alpha(h(x)) + \|L_F h(x)\| \tilde{\vartheta}. \tag{5.18}$$

Proof If $\tilde{\vartheta} \in \mathbb{R}_{\geq 0}$ is such that $\tilde{\vartheta} \leq \tilde{\vartheta}_{\max}$, then $\|L_F h(x)\| \tilde{\vartheta}_{\max} \geq \|L_F h(x)\| \tilde{\vartheta}$. Thus, if h is a RaCBF for (4.1), then for all $(x, \hat{\theta}) \in \mathbb{R}^n \times \Theta$

$$\sup_{u \in \mathcal{U}} \left\{ L_f h(x) + L_F h(x)\hat{\theta} + L_g h(x)u \right\} > -\alpha(h(x)) + \|L_F h(x)\| \tilde{\vartheta}_{\max}$$
$$\geq -\alpha(h(x)) + \|L_F h(x)\| \tilde{\vartheta}, \tag{5.19}$$

implying (5.18) holds. $\qquad \square$

The following assumption outlines the characteristics of parameter estimators that reduce the level of uncertainty online.

Assumption 5.2 There exists a parameter update law $\dot{\hat{\theta}} = \tau(\hat{\theta}, t)$, with τ locally Lipschitz in $\hat{\theta}$ and piecewise continuous in t such that $\hat{\theta}(t) \in \Theta$ for all $t \in \mathbb{R}_{\geq 0}$. Moreover, there exists a piecewise continuous and a non-increasing function $\tilde{\vartheta} : \mathbb{R}_{\geq 0} \to \mathbb{R}_{\geq 0}$ such that

$$\|\tilde{\theta}(t)\| \leq \tilde{\vartheta}(t) \leq \tilde{\vartheta}_{\max}, \quad \forall t \in \mathbb{R}_{\geq 0}. \tag{5.20}$$

Examples of parameter estimation routines satisfying the above assumption will be provided shortly. The following theorem constitutes the main result with regard to RaCBFs.

Theorem 5.2 *Let $h : \mathbb{R}^n \to \mathbb{R}$ be a RaCBF for (4.1) on a set C as in (3.3), and let Assumptions 5.1 and 5.2 hold. Define the set-valued map*

$$K(x, \hat{\theta}, \tilde{\vartheta}) := \left\{ u \in \mathcal{U} \mid L_f h(x) + L_F h(x)\hat{\theta} + L_g h(x)u \geq -\alpha(h(x)) + \|L_F h(x)\|\tilde{\vartheta} \right\}.$$

Then, any locally Lipschitz controller $u = k(x, \hat{\theta}, \tilde{\vartheta}(t))$ satisfying $k(x, \hat{\theta}, \tilde{\vartheta}(t)) \in K(x, \hat{\theta}, \tilde{\vartheta}(t))$ for all $(x, \hat{\theta}, \tilde{\vartheta}(t)) \in \mathbb{R}^n \times \Theta \times [0, \tilde{\vartheta}_{\max}]$ with $t \mapsto \tilde{\vartheta}(t)$ from Assumption 5.2, renders C forward invariant.

Proof The Lie derivative of h along the closed-loop vector field is

$$\begin{aligned}
\dot{h} &= L_f h(x) + L_F h(x)\theta + L_g h(x)k(x, \hat{\theta}, \tilde{\vartheta}(t)) \\
&= L_f h(x) + L_F h(x)\hat{\theta} + L_g h(x)k(x, \hat{\theta}, \tilde{\vartheta}(t)) + L_F h(x)\tilde{\theta}.
\end{aligned} \tag{5.21}$$

Now let $t \mapsto (x(t), \hat{\theta}(t))$ be the trajectories of the composite system

$$\begin{bmatrix} \dot{x} \\ \dot{\hat{\theta}} \end{bmatrix} = \begin{bmatrix} f(x) + F(x)\theta + g(x)k(x, \hat{\theta}, \tilde{\vartheta}(t)) \\ \tau(\hat{\theta}, t) \end{bmatrix},$$

whose existence and uniqueness on some maximal interval of existence are guaranteed given the assumptions on the controller and update law. Note that, as under Assumption 5.2 we have $\tilde{\vartheta}(t) \leq \tilde{\vartheta}_{\max}$, the set $K(x(t), \hat{\theta}(t), \tilde{\vartheta}(t))$ is non-empty for each t by Lemma 5.1. Hence, lower bounding $\dot{h}$ along the system trajectory yields

$$\begin{aligned}
\dot{h} &\geq L_f h(x(t)) + L_F h(x(t))\hat{\theta}(t) + L_g h(x(t))k(x(t), \hat{\theta}(t), \tilde{\vartheta}(t)) - \|L_F h(x(t))\|\|\tilde{\theta}(t)\| \\
&\geq -\alpha(h(x(t))) + \|L_F h(x(t))\|\tilde{\vartheta}(t) - \|L_F h(x(t))\|\|\tilde{\theta}(t)\| \\
&\geq -\alpha(h(x(t))).
\end{aligned} \tag{5.22}$$

As $\nabla h(x(t)) \neq 0$ for any $x(t) \in \partial C$ and $\dot{h} \geq -\alpha(h(x(t)))$, h is a barrier function for the closed-loop system and C is forward invariant by Theorem 3.2.   □

The above result demonstrates that combining a RaCBF with a parameter estimation routine that reduces the estimation error online allows for constructing controllers that robustly enforce safety while reducing conservatism as more data about the system becomes available. It is interesting to note that if the parameter estimation algorithm enforces convergence of the estimates to their true values, then the RaCBF controller converges to the standard CBF controller in the limit as time goes to infinity. It is also important to note that safety is guaranteed regardless of whether the parameter estimation error converges to zero—safety is guaranteed so long as the estimation error does not increase along a given trajectory. Given a RaCBF h and nominal adaptive controller k_0, inputs satisfying the RaCBF conditions can be enforced by solving the QP

$$k(x, \hat{\theta}, \tilde{\vartheta}) = \underset{u \in \mathcal{U}}{\arg\min} \quad \frac{1}{2}\|u - k_0(x, \hat{\theta})\|^2$$

$$\text{subject to} \quad L_f h(x) + L_F h(x)\hat{\theta} + L_g h(x)u \geq -\alpha(h(x)) + \|L_F h(x)\|\tilde{\vartheta},$$
$$(5.23)$$

which has a locally Lipschitz closed-form solution when $\mathcal{U} = \mathbb{R}^m$ given by

$$k(x, \hat{\theta}, \tilde{\vartheta}) = \begin{cases} k_0(x, \hat{\theta}) & \text{if } \psi(x, \hat{\theta}, \tilde{\vartheta}) \geq 0 \\ k_0(x, \hat{\theta}) - \frac{\psi(x, \hat{\theta}, \tilde{\vartheta})}{\|L_g h(x)^\top\|^2} L_g h(x)^\top & \text{if } \psi(x, \hat{\theta}, \tilde{\vartheta}) < 0, \end{cases} \quad (5.24)$$

where

$$\psi(x, \hat{\theta}, \tilde{\vartheta}) = L_f h(x) + L_F h(x)\hat{\theta} + L_g h(x)k_0(x, \hat{\theta}) + \alpha(h(x)) - \|L_F h(x)\|\tilde{\vartheta}.$$

Remark 5.1 If the nominal adaptive controller k_0 in (5.23) is generated by an aCLF or ES-aCLF, it may be necessary to use *different* estimates of θ in the objective function and constraint. That is, one may need to solve the QP

$$\min_{u \in \mathcal{U}} \quad \frac{1}{2}\|u - k_0(x, \hat{\theta}_{\text{clf}})\|^2$$

$$\text{subject to} \quad L_f h(x) + L_F h(x)\hat{\theta}_{\text{cbf}} + L_g h(x)u \geq -\alpha(h(x)) + \|L_F h(x)\|\tilde{\vartheta},$$

where $\hat{\theta}_{\text{cbf}}$ and $\hat{\theta}_{\text{clf}}$ are estimates generated by the update laws needed to guarantee safety and stability when using CBFs and CLFs, respectively. This redundancy in parameter estimation can be removed using the methods developed in Chap. 6.

We close this section by providing a particular example of a parameter estimator that satisfies Assumption 5.2 using the concurrent learning method introduced in Sect. 4.2. Recall that integrating (4.1) over some finite time interval $[t - \Delta t, t]$ yields the linear regression equation from (4.18)

$$\mathcal{Y}(t) = \mathcal{F}(t)\theta,$$

with $\mathcal{Y}$ and $\mathcal{F}$ defined as in (4.17). Using the above relation, one can update the parameters as

$$\dot{\hat{\theta}} = \gamma \sum_{j=1}^{M} \mathcal{F}_j(t)^\top \left[\mathcal{Y}_j(t) - \mathcal{F}_j(t)\hat{\theta} \right], \tag{5.25}$$

where $\gamma \in \mathbb{R}_{>0}$ is a learning gain, given a history stack $\mathcal{H} = \{(\mathcal{Y}_j, \mathcal{F}_j)\}_{j=1}^{M}$ consisting of tuples of $\mathcal{Y}$ and $\mathcal{F}$ recorded at various instances along the system trajectory.

Proposition 5.4 *Let the estimated parameters be updated according to (5.25) and consider the initial value problem*

$$\dot{\tilde{\vartheta}}(t) = -\gamma \lambda_{\min} \left(\sum_{j=1}^{M} \mathcal{F}_j(t)^\top \mathcal{F}_j(t) \right) \tilde{\vartheta}(t) \tag{5.26}$$

$$\tilde{\vartheta}(0) = \tilde{\vartheta}_{\max}.$$

Then, $\hat{\theta}(t) \in \Theta$ for all $t \in \mathbb{R}_{\geq 0}$, $t \mapsto \tilde{\vartheta}(t)$ is non-increasing, and $\|\tilde{\theta}(t)\| \leq \tilde{\vartheta}(t)$ for all $t \in \mathbb{R}_{\geq 0}$.

Proof Provided the parameters are updated according to (5.25), the parameter estimation error evolves according to

$$\dot{\tilde{\theta}} = -\gamma \sum_{j=1}^{M} \mathcal{F}_j(t)^\top \mathcal{F}_j(t)\tilde{\theta}, \tag{5.27}$$

which can be upper bounded as

$$\|\dot{\tilde{\theta}}\| \leq -\gamma \lambda_{\min} \left(\sum_{j=1}^{M} \mathcal{F}_j(t)^\top \mathcal{F}_j(t) \right) \|\tilde{\theta}\|. \tag{5.28}$$

As $\sum_{j=1}^{M} \mathcal{F}_j(t)^\top \mathcal{F}_j(t)$ is at least positive semidefinite for all $t \in \mathbb{R}_{\geq 0}$, we have $\|\dot{\tilde{\theta}}\| \leq 0$. By the same reasoning, we also have that $\dot{\tilde{\vartheta}} \leq 0$. Using the comparison lemma (Lemma 2.1) to solve the above differential inequality implies that

$$\|\tilde{\theta}(t)\| \leq \exp \left(-\gamma \int_0^t \lambda_{\min} \left(\sum_{j=1}^{M} \mathcal{F}_j(s)^\top \mathcal{F}_j(s) \right) ds \right) \|\tilde{\theta}(0)\|. \tag{5.29}$$

Furthermore, the solution to the initial value problem in (5.26) is given by

$$\tilde{\vartheta}(t) = \exp\left(-\gamma \int_0^t \lambda_{\min}\left(\sum_{j=1}^M \mathcal{F}_j(s)^\top \mathcal{F}_j(s)\right) ds\right)\tilde{\vartheta}_{\max}. \tag{5.30}$$

Since $\|\tilde{\theta}(0)\| \leq \tilde{\vartheta}_{\max}$, the above implies that $\|\tilde{\theta}(t)\| \leq \tilde{\vartheta}(t) \leq \tilde{\vartheta}_{\max}$ for all $t \in \mathbb{R}_{\geq 0}$, which, based on (5.12), also ensures $\hat{\theta}(t) \in \Theta$ for all $t \in \mathbb{R}_{\geq 0}$, as desired.   □

5.3 High Order Robust Adaptive Control Barrier Functions

In the previous section, we outlined a methodology for safe robust adaptive control using CBFs. The main limitation of such an approach is that it is contingent on knowledge of a CBF-like function, which implicitly requires knowledge of a controlled invariant set that can be expressed as the zero superlevel set of a single continuously differentiable function. As we discussed in Sect. 3.3, constructing such a function may be challenging as a user-defined state constraint set does not typically coincide with a controlled invariant set. Rather, one may need to search for a controlled invariant subset of the state constraint set—a process that, under certain assumptions, can be carried out using the high order CBF (HOCBF) approach.

In this section, we unite the HOCBF approach with the safe robust adaptive control approach outlined in the previous sections of this chapter. Similar to Sect. 3.3, we begin by considering the *state constraint set*

$$C_0 = \{x \in \mathbb{R}^n \mid h(x) \geq 0\},$$

where $h : \mathbb{R}^n \to \mathbb{R}$ has relative degree $r \in \mathbb{N}$. Our development proceeds by placing additional assumptions on the structure of the uncertainty in (4.1). Namely, we assume that if h has relative degree r, then the uncertain parameters only appear in the rth derivative of h along the system dynamics.

Assumption 5.3 If $h : \mathbb{R}^n \to \mathbb{R}$ has relative degree $r \in \mathbb{N}$ for (4.1), then $L_F L_f^i h(x) \neq 0$ for all $x \in \mathbb{R}^n$ and all $i \in \{0, 1, \ldots, r-2\}$, and there exists some nonempty set $\mathcal{R} \subset \mathbb{R}^n$ such that $L_F L_f^{r-1} h(x) \neq 0$ for all $x \in \mathcal{R}$.

Although we have only defined the notion of relative degree (see Def. 3.9) for control affine systems without uncertain parameters (2.10), the same criteria for relative degree applies to (4.1) as well. Following the same approach as in Sect. 3.3, we compute the derivative of h along the dynamics (4.1) until both the control input and uncertain parameters appear. To this end, we once again consider the collection of functions from (3.17):

$$\psi_0(x) = h(x)$$
$$\psi_i(x) = \dot{\psi}_{i-1}(x) + \alpha_i(\psi_{i-1}(x)), \quad \forall i \in \{1, \dots, r-1\},$$

where $\alpha_i \in \mathcal{K}_\infty^e$. If h has relative degree r and Assumption 5.3 holds, then each ψ_i, $i \in \{0, \dots, r-1\}$ will be independent of both u and θ for all $x \in \mathbb{R}^n$, whereas $\dot{\psi}_{r-1}(x, u, \theta)$ will depend on both u and θ. Each ψ_i, $i \in \{0, \dots, r-1\}$, is associated to a set $C_i \subset \mathbb{R}^n$ as in (3.18):

$$C_i = \{x \in \mathbb{R}^n \mid \psi_i(x) \geq 0\},$$

and we define the candidate safe set as in (3.19):

$$C = \bigcap_{i=0}^{r-1} C_i.$$

Before proceeding, we note that throughout this section it will be assumed that Assumption 5.1 holds so that there exists some maximum possible parameter estimation error $\tilde{\vartheta}_{\max}$. The following definition extends the concept of a HOCBF to nonlinear control systems with parametric uncertainty.

Definition 5.3 (*High order RaCBF*) Let $h : \mathbb{R}^n \to \mathbb{R}$ have relative degree $r \in \mathbb{N}$ for (4.1) such that $\nabla \psi_i(x) \neq 0$ for all $x \in \partial C_i$ for each $i \in \{0, \dots, r-1\}$, and let Assumption 5.3 hold. Then, h is said to be a *high order robust adaptive control barrier function* (HO-RaCBF) for (4.1) on a set C as in (3.19) if there exists $\alpha_r \in \mathcal{K}_\infty^e$ such that for all $(x, \hat{\theta}) \in \mathbb{R}^n \times \Theta$

$$\sup_{u \in \mathcal{U}} \left\{ L_f \psi_{r-1}(x) + L_F \psi_{r-1}(x)\hat{\theta} + L_g \psi_{r-1}(x)u \right\} \tag{5.31}$$
$$> -\alpha_r(\psi_{r-1}(x)) + \|L_F \psi_{r-1}(x)\|\tilde{\vartheta}_{\max}.$$

Similar to the previous section, h is a HO-RaCBF if $\mathcal{U} = \mathbb{R}^m$ and

$$\forall(x, \hat{\theta}) \in \mathbb{R}^n \times \Theta : L_g \psi_{r-1}(x) \implies L_f \psi_{r-1}(x) + L_F \psi_{r-1}(x)\hat{\theta}$$
$$> -\alpha_r(\psi_{r-1}(x)) + \|L_F \psi_{r-1}(x)\|\tilde{\vartheta}_{\max},$$

which can be further simplified when the parameters are matched.

Proposition 5.5 *Let h be a HOCBF for $\dot{x} = f(x) + g(x)u$ with $\mathcal{U} = \mathbb{R}^m$ on a set C defined as in (3.19), and suppose the parameters in (4.1) are matched. Then, h is a HO-RaCBF for (4.1) on C.*

Proof The proof follows the same steps as that of Proposition 4.1. □

Recall from Sect. 3.3 that it may be challenging to construct h such that the HOCBF conditions are satisfied at all points where $L_g \psi_{r-1}(x) = 0$; however, if such points are strictly bounded away from the boundary of the constraint set, h can always be modified so that the HOCBF conditions hold. Keeping in line with the approach taken in previous sections, we define the set-valued map

$$K_\psi(x, \hat\theta) = \{u \in \mathcal{U} \mid L_f \psi_{r-1}(x) + L_F \psi_{r-1}(x)\hat\theta + L_g \psi_{r-1}(x)u$$
$$\geq -\alpha_r(\psi_{r-1}(x)) + \|L_F \psi_{r-1}(x)\|\tilde\vartheta_{\max}\}, \tag{5.32}$$

that assigns to each $(x, \hat\theta) \in \mathbb{R}^n \times \Theta$ the set of controls satisfying the HO-RaCBF condition (5.31). The following proposition shows that drawing controllers (in a point-wise sense) from such a map renders C forward invariant for the closed-loop system

Proposition 5.6 *Let $h : \mathbb{R}^n \to \mathbb{R}$ be a HO-RaCBF for (4.1) on a set $C \subset \mathbb{R}^n$ as in (3.19) and let Assumption 5.1 hold. Then, any locally Lipschitz controller $u = k(x, \hat\theta)$ satisfying $k(x, \hat\theta) \in K_\psi(x, \hat\theta)$ for all $(x, \hat\theta) \in \mathbb{R}^n \times \Theta$ renders C forward invariant.*

Proof Define the closed-loop system vector field by

$$f_{\mathrm{cl}}(x, \hat\theta) := f(x) + F(x)\theta + g(x)k(x, \hat\theta),$$

and note that the Lie derivative of ψ_{r-1} along f_{cl} satisfies

$$\dot\psi_{r-1} = L_{f_{\mathrm{cl}}}(x, \hat\theta)$$
$$= L_f \psi_{r-1}(x) + L_F \psi_{r-1}(x)\theta + L_g \psi_{r-1}(x)k(x, \hat\theta)$$
$$= L_f \psi_{r-1}(x) + L_F \psi_{r-1}(x)\hat\theta + L_g \psi_{r-1}(x)k(x, \hat\theta) + L_F \psi_{r-1}(x)\tilde\theta$$
$$\geq L_f \psi_{r-1}(x) + L_F \psi_{r-1}(x)\hat\theta + L_g \psi_{r-1}(x)k(x, \hat\theta) - \|L_F \psi_{r-1}(x)\|\tilde\vartheta_{\max}$$
$$\geq -\alpha_r(\psi_{r-1}(x)).$$

The remainder of the proof then follows the same steps as those of Theorem 3.4. $\qquad\square$

Similar to RaCBFs, the validity of a HO-RaCBFs as a safety certificate remains valid as the level of uncertainty is reduced.

Lemma 5.2 *If h is a HO-RaCBF for (4.1) for a given $\tilde\vartheta_{\max} \in \mathbb{R}_{\geq 0}$, then it is also a HO-RaCBF for any $\tilde\vartheta \in [0, \tilde\vartheta_{\max}]$ in the sense that for all $(x, \hat\theta) \in \mathbb{R}^n \times \Theta$*

$$\sup_{u \in \mathcal{U}} \left\{ L_f \psi_{r-1}(x) + L_F \psi_{r-1}(x)\hat\theta + L_g \psi_{r-1}(x)u \right\}$$
$$> -\alpha_r(\psi_{r-1}(x)) + \|L_F \psi_{r-1}(x)\|\tilde\vartheta. \tag{5.33}$$

Proof Follows the same steps as that of Lemma 5.1. $\qquad\square$

Theorem 5.3 *Let* $h : \mathbb{R}^n \to \mathbb{R}$ *be a HO-RaCBF for (4.1) C as in (3.19), and let Assumptions 5.1 and 5.2 hold. Define the set-valued map*

$$K_\psi(x, \hat{\theta}, \tilde{\vartheta}) = \{u \in \mathcal{U} \mid L_f \psi_{r-1}(x) + L_F \psi_{r-1}(x)\hat{\theta} + L_g \psi_{r-1}(x)u$$
$$\geq -\alpha_r(\psi_{r-1}(x)) + \|L_F \psi_{r-1}(x)\|\tilde{\vartheta}\}.$$

Then, any locally Lipschitz controller $u = k(x, \hat{\theta}, \tilde{\vartheta}(t))$ *satisfying* $k(x, \hat{\theta}, \tilde{\vartheta}(t)) \in K(x, \hat{\theta}, \tilde{\vartheta}(t))$ *for all* $(x, \hat{\theta}, \tilde{\vartheta}(t)) \in \mathbb{R}^n \times \Theta \times [0, \tilde{\vartheta}_{\max}]$ *with* $t \mapsto \tilde{\vartheta}(t)$ *from Assumption 5.2, renders C forward invariant.*

Proof The proof is a combination of that of Theorem 5.2 and Proposition 5.6. □

Given a HO-RaCBF and a parameter estimator satisfying Assumption 5.2, inputs satisfying the conditions of the above theorem can be computed by solving the QP

$$k(x, \hat{\theta}, \tilde{\vartheta}) = \underset{u \in \mathcal{U}}{\arg\min} \quad \frac{1}{2}\|u - k_0(x, \hat{\theta})\|^2$$
$$\text{subject to} \quad L_f \psi_{r-1}(x) + L_F \psi_{r-1}(x)\hat{\theta} + L_g \psi_{r-1}(x)u \tag{5.34}$$
$$\geq -\alpha_r(\psi_{r-1}(x)) + \|L_F \psi_{r-1}(x)\|\tilde{\vartheta},$$

where k_0 is a locally Lipschitz nominal adaptive controller. When $\mathcal{U} = \mathbb{R}^m$, the closed-form solution to the above QP is given by

$$k(x, \hat{\theta}, \tilde{\vartheta}) = \begin{cases} k_0(x, \hat{\theta}), & \text{if } \Psi(x, \hat{\theta}, \tilde{\vartheta}) \geq 0 \\ k_0(x, \hat{\theta}) - \frac{\Psi(x, \hat{\theta}, \tilde{\vartheta})}{\|L_g \psi_{r-1}(x)^\top\|^2} L_g \psi_{r-1}(x)^\top, & \text{if } \Psi(x, \hat{\theta}, \tilde{\vartheta}) < 0, \end{cases} \tag{5.35}$$

where

$$\Psi(x, \hat{\theta}, \tilde{\vartheta}) := L_f \psi_{r-1}(x) + L_F \psi_{r-1}(x)\hat{\theta} + L_g \psi_{r-1}(x)k_0(x, \hat{\theta})$$
$$+ \alpha_r(\psi_{r-1}(x)) - \|L_F \psi_{r-1}(x)\|\tilde{\vartheta},$$

and is locally Lipschitz.

5.4 Numerical Examples

Example 5.1 We illustrate many of the ideas introduced in this chapter using a more complex version of the robot motion planning problem introduced in Example 3.2. In particular, we now consider a mobile robot modeled as a planar double integrator with uncertain friction effects

$$\ddot{q} = u - \mu\dot{q}, \tag{5.36}$$

where $q \in \mathbb{R}^2$ denotes the position of the robot, $u \in \mathbb{R}^2$ is its commanded acceleration, and $\mu \in \mathbb{R}^2$ is a vector of uncertain friction coefficients. Taking the state as $x := [q^\top \; \dot{q}^\top]^\top$ and the uncertain parameters as $\theta := \mu$ allows the system to be expressed in the form of (4.1) as

$$\begin{bmatrix} \dot{x}_1 \\ \dot{x}_2 \\ \dot{x}_3 \\ \dot{x}_4 \end{bmatrix} = \begin{bmatrix} x_3 \\ x_4 \\ 0 \\ 0 \end{bmatrix} + \begin{bmatrix} 0 & 0 \\ 0 & 0 \\ -x_3 & 0 \\ 0 & -x_4 \end{bmatrix} \theta + \begin{bmatrix} 0 & 0 \\ 0 & 0 \\ 1 & 0 \\ 0 & 1 \end{bmatrix} u. \tag{5.37}$$

For simplicity, we set $\theta = [1 \; 1]^\top$. Note from (5.36) that the uncertain parameters are clearly matched to the control input. The objective is to stabilize the system to the origin while avoiding a set of static obstacle in the workspace. To achieve the stabilization objective, we construct an exponentially stabilizing adaptive control Lyapunov function (ES-aCLF) introduced in Chap. 4 by solving an LQR problem for the nominal system without any uncertainty. The safety objective is achieved by considering a collection of state constraint sets of the form (3.16) with

$$h_i(x) = (x_1 - y_{1,i})^2 + (x_2 - y_{2,i})^2 - R_i^2,$$

where $[y_{1,i} \; y_{2,i}]^\top \in \mathbb{R}^2$ denote the location of the obstacle's center and $R_i \in \mathbb{R}_{>0}$ denotes its radius, which are used to form a collection of candidate HO-RaCBFs defined by the extended class $\mathcal{K}_\infty$ functions $\alpha_1(r) = \alpha_2(r) = r$. With straightforward calculations one can verify that such a function has relative degree 2 for the control input and uncertain parameters. To demonstrate the impact of reducing the level of uncertainty online, we simulate the system under the resulting HO-RaCBF/ES-aCLF controller and compare the results to controllers that account for the worst-case uncertainty without any online adaptation. To learn the uncertain parameters online, we use the concurrent learning estimator outlined in (5.25), where we maintain a history stack with $M = 20$ entries using an integration window of $\Delta t = 0.5$ s and a learning gain of $\gamma = 10$. As noted in Remark 5.1, different update laws are required for the ES-aCLF and HO-RaCBF to provide stability and safety guarantees, respectively. Hence, we maintain two separate estimates of the uncertain parameters, where the parameters for the ES-aCLF controller are updated using the estimator proposed in Theorem 4.4 with $\Gamma = I_{2\times2}$ and $\gamma = 10$. Note that, although two different update laws must be used, the same history stack can be shared between them.

To demonstrate the relationship between adaptation and safety, we run a set of simulations comparing performance using the HO-RaCBF controller to that under a purely robust approach (i.e., accounting for the maximum possible parameter estimation error without adaption), where the set of possible parameters Θ satisfying Assumption 5.1 are varied. The results of these simulations are reported in Figs. 5.1, 5.2 and 5.3. As shown in Figures 5.1 and 5.2, the trajectories under both controllers are safe for all levels of uncertainty; however, for larger levels of uncertainty, the purely robust controller is overly conservative, causing the system trajectory to diverge from the origin. In contrast, the adaptive controller reduces

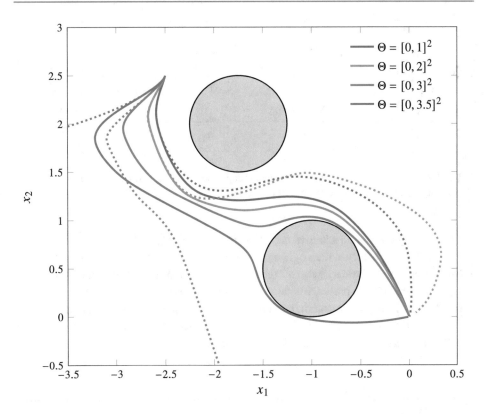

Fig. 5.1 Position trajectory for the mobile robot under each controller across four different uncertainty sets, where the solid lines denote trajectories under the adaptive controller and the dashed lines denote trajectories under the purely robust controller. The gray disks represent obstacles in the workspace

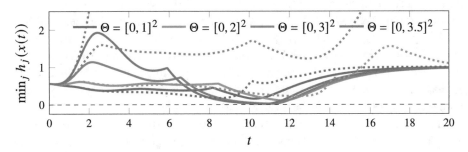

Fig. 5.2 Minimum value among the two HO-RaCBFs point-wise in time along each system trajectory. The solid and dashed curves have the same interpretation as those in Fig. 5.1 and the dashed black line denotes $h(x) = 0$

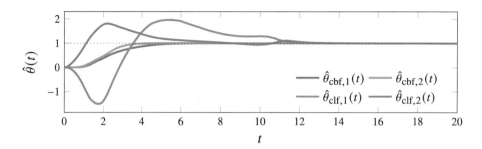

Fig. 5.3 Estimates of the uncertain parameters used in the ES-aCLF and HO-RaCBF controller for the simulation corresponding to the uncertainty set $\Theta = [0, 3]^2$

the uncertainty online and achieves dual objectives of stability and safety. In fact, the convergence of the trajectory to the origin under the adaptive controller is minimally affected by the initial level of uncertainty, whereas the trajectory under the robust controller fails to converge to the origin in the presence of large uncertainty. The ability of the concurrent learning estimators to identify the uncertain parameters is demonstrated in Fig. 5.3, which shows the trajectory of the estimated parameters used in the ES-aCLF and HO-RaCBF controller, both of which converge to their true values in just under 15 s.

5.5   Notes

In this chapter, we discussed extensions of traditional nonlinear adaptive control techniques from stability specifications to safety specifications. Early works on incorporating safety constraints into adaptive control approaches relied on the use of barrier Lyapunov functions (BLF) [1]; however, as noted in Chap. 3 BLFs are often overly restrictive. Other promising approaches to safe adaptive control leverage model predictive control (MPC). In this setting, safety in the presence of uncertainties is typically enforced using a tube-based MPC approach, in which the size of the tube is reduced by reducing the uncertainty in the parameter estimates using various parameter estimation routines [2–5].

The notion of an *adaptive control barrier function* (aCBF) was introduced in [6], and extended the adaptive control Lyapunov function (aCLF) paradigm [7] from stability to safety problems. These ideas were extended using the notion of a *robust aCBF* (RaCBF) in [8], where techniques based on set membership identification [9] were used to reduce the level of uncertainty in the parameter estimates online. The idea of using concurrent learning to reduce parameter uncertainty when using CBFs was introduced in [10], with related works [11] also exploiting the concurrent learning paradigm to reduce parameter uncertainty online. The extension of RaCBFs to high relative degree safety constraints was introduced in [12]. Other related works uniting ideas from adaptive control and CBFs include [13–16]. A different version of an aCBF was introduced in [17], where adaptation was used to adhere to potentially time-varying control bounds rather than to estimate unknown parameters online.

Beyond traditional adaptive control methods, other learning-based approaches have also been combined with CBFs to develop safety-critical controllers for uncertain systems. The methods in [18–20] leverage an episodic learning framework to reduce the impact of uncertainties on the CBF conditions, allowing potential safety violations to be reduced episodically as more data about the system is collected. By delegating the learning process to offline computations, such approaches allow for the use of powerful functions approximators, such as deep neural networks, to represent unknown terms in the system dynamics. Along similar lines, probabilistic approaches that leverage Gaussian processes (GPs) as function approximators have been combined with CBFs as they allow for making high confidence statements regarding estimated model errors. Works that combine GPs with CBFs to develop controllers for uncertain systems with probabilistic safety guarantees include [21–24].

References

1. Tee KP, Ge SS, Tay EH (2009) Barrier lyapunov functions for the control of output-constrained nonlinear systems. Automatica 45(4):918–927
2. Tanaskovic M, Fagiano L, Smith R, Morari M (2014) Adaptive receding horizon control for constrained mimo systems. Automatica 50:3019–3029
3. Lopez BT (2019) Adaptive robust model predictive control for nonlinear systems. PhD thesis, Massachusetts Institute of Technology
4. Lu X, Cannon M, Koksal-Rivet D (2021) Robust adaptive model predictive control: performance and parameter estimation. Int J Robust Nonlinear Control 31(18):8703–8724
5. Köhler J, Kötting P, Soloperto R, Allgöwer F, Müller MA (2021) A robust adaptive model predictive control framework for nonlinear uncertain systems. Int J Robust Nonlinear Control 31(18)
6. Taylor AJ, Ames AD (2020) Adaptive safety with control barrier functions. In: Proceedings of the American control conference, pp 1399–1405
7. Krstić M, Kokotović P (1995) Control lyapunov functions for adaptive nonlinear stabilization. Syst & Control Lett 26(1):17–23
8. Lopez BT, Slotine JJ, How JP (2021) Robust adaptive control barrier functions: an adaptive and data-driven approach to safety. IEEE Control Syst Lett 5(3):1031–1036
9. Kosut RL, Lau MK, Boyd SP (1992) Set-membership identification of systems with parametric and nonparametric uncertainty. IEEE Trans Autom Control 37(7):929–941
10. Isaly A, Patil OS, Sanfelice RG, Dixon WE (2021) Adaptive safety with multiple barrier functions using integral concurrent learning. In: Proceedings of the American control conference, pp 3719–3724
11. Azimi V, Hutchinson S (2021) Exponential control lyapunov-barrier function using a filtering-based concurrent learning adaptive approach. IEEE Trans Autom Control
12. Cohen MH, Belta C (2022) High order robust adaptive control barrier functions and exponentially stabilizing adaptive control lyapunov functions. In: Proceedings of the American control conference pp 2233–2238

13. Zhao P, Mao Y, Tao C, Hovakimyan N, Wang X (2020) Adaptive robust quadratic programs using control lyapunov and barrier functions. In: Proceedings of the IEEE conference on decision and control, pp 3353–3358
14. Black M, Arabi E, Panagou D (2021) A fixed-time stable adaptation law for safety-critical control under parametric uncertainty. In: Proceedings of the European control conference, pp 1328–1333
15. Maghenem M, Taylor AJ, Ames AD, Sanfelice RG (2021) Adaptive safety using control barrier functions and hybrid adaptation. In: Proceedings of the American control conference, pp 2418–2423
16. Nguyen Q, Sreenath K (2022) L1 adaptive control barrier functions for nonlinear underactuated systems. In: Proceedings of the American control conference, pp 721–728
17. Xiao W, Belta C, Cassandras CG (2022) Adaptive control barrier functions. IEEE Trans Autom Control 67(5):2267–2281
18. Taylor AJ, Singletary A, Yue Y, Ames A (2020) Learning for safety-critical control with control barrier functions. In: Proceedings of the 2nd annual conference on learning for dynamics and control. Proceedings of machine learning research, vol 120, pp 708–717
19. Taylor AJ, Singletary A, Yue Y, Ames AD (2020) A control barrier perspective on episodic learning via projection-to-state safety. IEEE Control Syst Lett 5(3):1019–1024
20. Csomay-Shanklin N, Cosner RK, Dai M, Taylor AJ, Ames AD (2021) Episodic learning for safe bipedal locomotion with control barrier functions and projection-to-state safety. In: Proceedings of the 3rd annual conference on learning for dynamics and control. Proceedings of machine learning research, vol 144, pp 1041–1053
21. Castaneda F, Choi JJ, Zhang B, Tomlin CJ, Sreenath K (2021) Pointwise feasibility of gaussian process-based safety-critical control under model uncertainty. In: Proceedings of the IEEE conference on decision and control, pp 6762–6769
22. Dhiman V, Khojasteh MJ, Franceschetti M, Atanasov N (2021) Control barriers in bayesian learning of system dynamics. IEEE Trans Autom Control
23. Fan DD, Nguyen J, Thakker R, Alatur N, Agha-mohammadi A, Theodorou EA (2020) Bayesian learning-based adaptive control for safety critical systems. In: Proceedings of the IEEE international conference on robotics and automation, pp 4093–4099
24. Cheng R, Khojasteh MJ, Ames AD, Burdick JW (2020) Safe multi-agent interaction through robust control barrier functions with learned uncertainties. In: Proceedings of the IEEE conference on decision and control, pp 777–783

A Modular Approach to Adaptive Safety-Critical Control

In the previous chapter, we introduced adaptive control barrier functions (aCBFs) for systems with uncertain parameters. Central to that approach was the construction of a suitable parameter estimation algorithm that continuously reduced the level of uncertainty in the parameter estimates using data collected online. In this chapter, by unifying the concepts of *input-to-state stability* (ISS) and *input-to-state-safety* (ISSf), we develop a framework for modular adaptive control that addresses some limitations of that method. Specifically, we show how to allow more freedom in the parameter estimation algorithm, how to relax the required knowledge on the parameter bounds, and how to reduce the redundancy in parameter estimation necessary for safety and stability. In Sect. 6.1, we introduce input-to-state stability (ISS). The concept of modular adaptive stabilization is defined in Sect. 6.2. The ISS concept is extended to input-to-state safety (ISSf) in Sect. 6.3. We include numerical examples in Sect. 6.4 and conclude with final remarks, references, and suggestions for further reading in Sect. 6.5.

As in the previous few chapters,[1] our central object of interest is the nonlinear dynamical system with parametric uncertainty (4.1), given here again for convenience:

$$\dot{x} = f(x) + F(x)\theta + g(x)u,$$

where $x \in \mathbb{R}^n$ is the state, $u \in \mathcal{U} \subset \mathbb{R}^m$ is the control input, $\theta \in \mathbb{R}^p$ are the uncertain parameters, and $f : \mathbb{R}^n \to \mathbb{R}^n$, $F : \mathbb{R}^n \to \mathbb{R}^{n \times p}$, $g : \mathbb{R}^n \to \mathbb{R}^{n \times m}$ characterize the system dynamics. Letting $\hat{\theta} \in \mathbb{R}^p$ be an estimate of the uncertain parameters, recall that the

---

[1] The term modular adaptive control is often synonymous with indirect adaptive control or estimation-based adaptive control.

© The Author(s), under exclusive license to Springer Nature Switzerland AG 2023
M. Cohen and C. Belta, *Adaptive and Learning-Based Control of Safety-Critical Systems*,
Synthesis Lectures on Computer Science,
https://doi.org/10.1007/978-3-031-29310-8_6

parameter estimation error $\tilde{\theta} \in \mathbb{R}^p$ is defined as

$$\tilde{\theta} = \theta - \hat{\theta}.$$

The estimation error allows (4.1) to be equivalently represented as

$$\dot{x} = f(x) + F(x)\hat{\theta} + g(x)u + F(x)\tilde{\theta}. \tag{6.1}$$

The approach taken in this chapter is to view the term $F(x)\tilde{\theta}$ as a disturbance input to the nominal system dynamics $f(x) + F(x)\hat{\theta} + g(x)u$, and then to characterize the impact of such a disturbance on the stability and safety properties of the nominal system using the notions of ISS and ISSf.

6.1   Input-to-State Stability

We begin our development by reviewing the concept of input-to-state stability (ISS), which allows for characterizing stability of nonlinear systems in the presence of disturbances/uncertainties. In this section, we briefly introduce the notion of ISS for the general disturbed dynamical system

$$\dot{x} = f(x, d), \tag{6.2}$$

where $x \in \mathbb{R}^n$ is the system state, $d \in \mathbb{R}^p$ is a disturbance input, and $f : \mathbb{R}^n \times \mathbb{R}^p \to \mathbb{R}^n$ is a vector field, locally Lipschitz in its arguments, and later specialize this definition to our system of interest.

Definition 6.1 (*Input-to-state stability*) System (6.2) is said to be *input-to-state stable* if there exists $\beta \in \mathcal{KL}$ and $\iota \in \mathcal{K}_\infty$ such that for any initial condition $x_0 \in \mathbb{R}^n$ and any continuous disturbance signal $d(\cdot)$ the solution $t \mapsto x(t)$ of the initial value problem

$$\dot{x}(t) = f(x(t), d(t))$$
$$x(0) = x_0,$$

satisfies

$$\|x(t)\| \leq \beta(\|x_0\|, t) + \iota(\|d\|_\infty), \tag{6.3}$$

where $\|d\|_\infty := \sup_{t \in \mathbb{R}_{\geq 0}} \|d(t)\|$. If $\beta(r, s) = cr \exp(-\lambda s)$ for some $c, \lambda \in \mathbb{R}_{>0}$, then (6.2) is said to be *exponentially ISS* (eISS).

The definition of ISS implies that trajectories remain bounded and will converge to a ball about the origin, the size of which depends on the magnitude of the disturbance input. Clearly, in the absence of disturbances (i.e., if $\|d\|_\infty = 0$) we recover the definition of asymptotic stability. As with the definitions of stability introduced in Chap. 2, verifying the ISS property using Definition 6.1 requires knowledge of the system trajectories, which are difficult to obtain for nonlinear systems in general. Fortunately, ISS properties can be given a Lyapunov-like characterization using the notion of an *ISS Lyapunov function*.

Definition 6.2 (*ISS Lyapunov function*) A continuously differentiable function $V : \mathbb{R}^n \to \mathbb{R}_{\geq 0}$ is said to be an *ISS Lyapunov function* for (6.2) if there exist $\alpha_1, \alpha_2, \rho \in \mathcal{K}_\infty$ and $\alpha_3 \in \mathcal{K}$ such that, for all $(x, d) \in \mathbb{R}^n \times \mathbb{R}^p$,

$$\alpha_1(\|x\|) \leq V(x) \leq \alpha_2(\|x\|), \tag{6.4}$$

$$\|x\| \geq \rho(\|d\|) \implies L_f V(x, d) \leq -\alpha_3(\|x\|). \tag{6.5}$$

The following theorem shows that the existence of an ISS Lyapunov function is sufficient to establish ISS of (6.2).

Theorem 6.1 *Let V be an ISS Lyapunov function for (6.2). Then, (6.2) is ISS as in (6.3) with $\iota(r) = (\alpha_1^{-1} \circ \alpha_2 \circ \rho)(r)$.*

We now specialize the notion of ISS to the uncertain control system (6.1) with the parameter estimation error acting as a disturbance input to the nominal dynamics.

Definition 6.3 (*Exponential ISS-CLF*) A continuously differentiable function $V : \mathbb{R}^n \to \mathbb{R}_{\geq 0}$ is said to be an *exponential input-to-state stable control Lyapunov function* (eISS-CLF) for (6.1) if there exist positive constants $c_1, c_2, c_3, \varepsilon \in \mathbb{R}_{>0}$ such that

$$c_1 \|x\|^2 \leq V(x) \leq c_2 \|x\|^2, \quad \forall x \in \mathbb{R}^n, \tag{6.6}$$

and for all $(x, \hat{\theta}) \in (\mathbb{R}^n \setminus \{0\}) \times \mathbb{R}^p$,

$$\inf_{u \in \mathcal{U}} \left\{ L_f V(x) + L_F V(x)\hat{\theta} + L_g V(x)u \right\} < -c_3 V(x) - \tfrac{1}{\varepsilon} \|L_F V(x)\|^2. \tag{6.7}$$

Similar to the CLFs from earlier chapters, the above definition allows to construct the set-valued map

$$K_{\text{ISS}}(x, \hat{\theta}) := \{u \in \mathcal{U} \mid L_f V(x) + L_F V(x)\hat{\theta} + L_g V(x)u \leq -c_3 V(x) - \tfrac{1}{\varepsilon} \|L_F V(x)\|^2\} \tag{6.8}$$

that assigns to each $(x, \hat{\theta}) \in \mathbb{R}^n \times \mathbb{R}^p$ a set of control values satisfying the conditions of Definition 6.3. The following theorem shows that any locally Lipschitz controller belonging to the above set enforces ISS of (6.1).

Theorem 6.2 *Let V be an eISS-CLF for (6.1) and assume that $\tilde{\theta}(\cdot) \in \mathcal{L}_\infty$. Then, any controller $u = k(x, \hat{\theta}) \in K_{\text{ISS}}(x, \hat{\theta})$ locally Lipschitz on $(x, \hat{\theta}) \in (\mathbb{R}^n \setminus \{0\}) \times \mathbb{R}^p$ renders (6.1) eISS in the sense that, for all $t \in \mathbb{R}_{\geq 0}$,*

$$\underbrace{\|x(t)\| \leq \sqrt{\frac{c_2}{c_1}} \|x(0)\| \exp(-\tfrac{1}{2}c_3 t)}_{\beta(\|x(0)\|, t)} + \underbrace{\frac{1}{2}\sqrt{\frac{\varepsilon}{c_1 c_3}} \|\tilde{\theta}\|_\infty}_{\iota(\|\tilde{\theta}\|_\infty)}.$$ (6.9)

Proof The Lie derivative of V along the closed-loop dynamics can be bounded as

$$\begin{aligned}
\dot{V} &= L_f V(x) + L_F V(x)\hat{\theta} + L_g V(x)k(x,\hat{\theta}) + L_F V(x)\tilde{\theta} \\
&\leq -c_3 V(x) - \tfrac{1}{\varepsilon}\|L_F V(x)\|^2 + L_F V(x)\tilde{\theta} \\
&\leq -c_3 V(x) + \tfrac{\varepsilon}{4}\|\tilde{\theta}\|^2 \\
&\leq -c_3 V(x) + \tfrac{\varepsilon}{4}\|\tilde{\theta}\|_\infty^2,
\end{aligned}$$ (6.10)

where the first inequality follows from the definition of K_{ISS}, the second from completing squares, and the third from the assumption that $\tilde{\theta}(\cdot) \in \mathcal{L}_\infty$. Invoking the comparison lemma (Lemma 2.1) and using (6.6) yields

$$\begin{aligned}
\|x(t)\| &\leq \sqrt{\frac{c_2}{c_1}\|x(0)\|^2 \exp(-c_3 t) + \frac{\varepsilon}{4 c_1 c_3}\|\tilde{\theta}\|_\infty^2} \\
&\leq \sqrt{\frac{c_2}{c_1}}\|x(0)\| \exp(-\tfrac{1}{2}c_3 t) + \frac{1}{2}\sqrt{\frac{\varepsilon}{c_1 c_3}}\|\tilde{\theta}\|_\infty,
\end{aligned}$$ (6.11)

where the second inequality follows from the fact that the square root is a subadditive function, which implies that (6.1) is eISS as in (6.9), as desired. ∎

Similar to previous chapters, given an eISS-CLF, control inputs satisfying the above theorem can be computed for any $(x, \hat{\theta})$ by solving the optimization problem

$$k(x, \hat{\theta}) = \arg\min_{u \in \mathcal{U}} \quad \frac{1}{2}\|u\|^2$$

$$\text{subject to} \quad L_f V(x) + L_F V(x)\hat{\theta} + L_g V(x)u \leq -c_3 V(x) - \tfrac{1}{\varepsilon}\|L_F V(x)\|^2,$$ (6.12)

which is a QP when $\mathcal{U} = \mathbb{R}^m$ or $\mathcal{U}$ is a convex polytope. If $\mathcal{U} = \mathbb{R}^m$, the closed-form expression to (6.12) is given by

$$k(x, \hat{\theta}) = \begin{cases} 0, & \text{if } \psi(x, \hat{\theta}) \leq 0 \\ -\dfrac{\psi(x,\hat{\theta})}{\|L_g V(x)\|^2} L_g V(x)^\top, & \text{if } \psi(x, \hat{\theta}) > 0, \end{cases}$$ (6.13)

where

$$\psi(x, \hat{\theta}) := L_f V(x) + L_F V(x)\hat{\theta} + c_3 V(x) + \tfrac{1}{\varepsilon}\|L_F V(x)\|^2.$$

Note that when $\mathcal{U} = \mathbb{R}^m$ and the parameters in (4.1) are matched, the construction of an eISS-CLF can be done similarly to the construction of an aCLF (i.e., independently of the uncertain parameters). This follows from the fact that the statement that V is an eISS-CLF when $\mathcal{U} = \mathbb{R}^m$ is equivalent to the statement that

$$L_g V(x) = 0 \implies L_f V(x) + L_F V(x)\hat{\theta} < -c_3 V(x) - \tfrac{1}{\varepsilon}\|L_F V(x)\|^2.$$

Hence, when the parameters are matched $L_g V(x) = 0 \implies L_F V(x) = 0$, and the above condition reduces to the standard ES-CLF condition for the nominal dynamics. This observation is summarized in the following proposition.

Proposition 6.1 *Let $V : \mathbb{R}^n \to \mathbb{R}_{\geq 0}$ be an ES-CLF for $\dot{x} = f(x) + g(x)u$ with $\mathcal{U} = \mathbb{R}^m$. If the parameters in (6.1) are matched, then V is an eISS-CLF for (6.1) with $\mathcal{U} = \mathbb{R}^m$.*

6.2   Modular Adaptive Stabilization

In the previous section we demonstrated how the notion of ISS can be used to characterize the impact of parameter estimation error on the stability of an adaptive control system, and how eISS-CLFs could be used to construct controllers ensuring ISS of the closed-loop system. In the present section we outline properties of a general class of parameter estimators that can be combined with eISS-CLF-based controllers to provide asymptotic stability guarantees, rather than ISS guarantees. The following lemma outlines the characteristics of such a class of parameter estimators.

Lemma 6.1 *Consider a parameter update law $\dot{\hat{\theta}} = \tau(\hat{\theta}, t)$, with τ locally Lipschitz in its first argument and piecewise continuous in its second, and a Lyapunov-like function $V_\theta : \mathbb{R}^p \times \mathbb{R}_{\geq 0} \to \mathbb{R}_{\geq 0}$ satisfying*

$$\eta_1 \|\tilde{\theta}\|^2 \leq V_\theta(\tilde{\theta}, t) \leq \eta_2 \|\tilde{\theta}\|^2, \quad \forall(\tilde{\theta}, t) \in \mathbb{R}^p \times \mathbb{R}_{\geq 0}, \tag{6.14}$$

for some $\eta_1, \eta_2 \in \mathbb{R}_{>0}$. Provided

$$\dot{V}_\theta(\tilde{\theta}, t) \leq 0, \quad \forall(\tilde{\theta}, t) \in \mathbb{R}^p \times \mathbb{R}_{\geq 0}, \tag{6.15}$$

then $\tilde{\theta}(\cdot) \in \mathcal{L}_\infty$. Furthermore, if there exists a pair $(\eta_3, T) \in \mathbb{R}_{>0} \times \mathbb{R}_{\geq 0}$ such that

$$\dot{V}_\theta(\tilde{\theta}, t) \leq -\eta_3 \|\tilde{\theta}\|^2, \quad \forall(\tilde{\theta}, t) \in \mathbb{R}^p \times \mathbb{R}_{\geq T}, \tag{6.16}$$

then $\tilde{\theta}(\cdot) \in \mathcal{L}_2 \cap \mathcal{L}_\infty$ and

$$\|\tilde{\theta}(t)\| \leq \frac{\eta_2}{\eta_1} \|\tilde{\theta}(0)\| e^{\frac{\eta_3}{2\eta_2}T} e^{-\frac{\eta_3}{2\eta_2}t}, \quad \forall t \in \mathbb{R}_{\geq 0}. \tag{6.17}$$

Proof Since $\dot{V}_\theta(\tilde{\theta}(t), t) \leq 0$ for all $t \in \mathbb{R}_{\geq 0}$, $V_\theta(\tilde{\theta}(t), t)$ is nonincreasing and $V_\theta(\tilde{\theta}(t), t) \leq V_\theta(\tilde{\theta}(0), 0)$ for all $t \in \mathbb{R}_{\geq 0}$. Using (6.14) this implies that for all $t \in \mathbb{R}_{\geq 0}$

$$\|\tilde{\theta}(t)\| \le \sqrt{\tfrac{\eta_2}{\eta_1}}\|\tilde{\theta}(0)\|, \tag{6.18}$$

and thus $\tilde{\theta}(\cdot) \in \mathcal{L}_\infty$. For $t \in \mathbb{R}_{\geq T}$, $\dot{V}_\theta(\tilde{\theta}(t), t) \le -\eta_3\|\tilde{\theta}(t)\|^2$, which, after using the comparison lemma and (6.14), implies

$$
\begin{aligned}
\|\tilde{\theta}(t)\| &\le \sqrt{\tfrac{\eta_2}{\eta_1}}\|\tilde{\theta}(T)\|e^{-\frac{\eta_3}{2\eta_2}(t-T)} \\
&\le \tfrac{\eta_2}{\eta_1}\|\tilde{\theta}(0)\|e^{\frac{\eta_3}{2\eta_2}T}e^{-\frac{\eta_3}{2\eta_2}t},
\end{aligned}
\tag{6.19}
$$

for all $t \in \mathbb{R}_{\geq T}$. This bound is also valid for all $t \in \mathbb{R}_{\geq 0}$ as stated in (6.17) since $1 \le \exp(-\frac{\eta_3}{2\eta_2}(t - T))$ for all $t \in [0, T]$. The bound in (6.17) also implies $\tilde{\theta}(\cdot) \in \mathcal{L}_2$ since

$$
\begin{aligned}
\int_0^t \|\tilde{\theta}(s)\|^2 ds &\le (\tfrac{\eta_2}{\eta_1}\|\tilde{\theta}(0)\|e^{\frac{\eta_3}{2\eta_2}T})^2 \int_0^t e^{-\frac{\eta_3}{\eta_2}s}ds \\
&= (\tfrac{\eta_2}{\eta_1}\|\tilde{\theta}(0)\|e^{\frac{\eta_3}{2\eta_2}T})^2(-\tfrac{\eta_2}{\eta_3}(e^{-\frac{\eta_3}{\eta_2}t} - 1)),
\end{aligned}
$$

which, after taking limits as $t \to \infty$, implies that

$$\int_0^\infty \|\tilde{\theta}(s)\|^2 ds \le \tfrac{\eta_2^3}{\eta_3\eta_1^2}\|\tilde{\theta}(0)\|^2 e^{\frac{\eta_3}{\eta_2}T} < \infty. \tag{6.20}$$

Combining (6.20) and (6.18) yields $\tilde{\theta}(\cdot) \in \mathcal{L}_2 \cap \mathcal{L}_\infty$. $\qquad\square$

The condition in (6.15) requires that the parameter estimation error remains bounded for all time - a property satisfied by a variety of standard estimation algorithms (e.g., gradient descent, recursive least squares, etc.). The condition in (6.16) is more restrictive. It asks, after a certain time period, for the parameter estimates to exponentially converge to their true values, and is reminiscent of the concurrent learning parameter estimators introduced in earlier chapters. Later in this chapter we will provide specific examples of concurrent learning-based parameter estimators satisfying the conditions of Lemma 6.1. We show in the following theorem that combining a parameter estimator satisfying the conditions of Lemma 6.1 with a controller satisfying the conditions in (6.7) renders the origin of the closed-loop system asymptotically stable.

Theorem 6.3 *If V is an eISS-CLF for (6.1) and the conditions of Lemma 6.1 hold, then any bounded controller $u = k(x, \hat{\theta})$ locally Lipschitz on $(\mathbb{R}^n \setminus \{0\}) \times \mathbb{R}^p$ satisfying $k(x, \hat{\theta}) \in K_{ISS}(x, \hat{\theta})$ for all $(x, \hat{\theta}) \in \mathbb{R}^n \times \mathbb{R}^p$ renders the origin of (6.1) asymptotically stable.*

To prove the above theorem we require a classical tool used extensively in adaptive control known as *Barbalat's Lemma*.

Lemma 6.2 (Barbalat's Lemma) *Consider a signal $x(\cdot)$ and suppose that $x(\cdot), \dot{x}(\cdot) \in \mathcal{L}_\infty$ and $x(\cdot) \in \mathcal{L}_2$. Then $\lim_{t \to \infty} x(t) = 0$.*

Proof (of Theorem 6.3) The stability of the origin follows directly from Theorem 6.2 since $\tilde{\theta}(\cdot) \in \mathcal{L}_\infty$ by Lemma 6.1. To show that the origin is also attractive in the sense that $\lim_{t \to \infty} x(t) = 0$ we rearrange the third line of (6.10):

$$c_1 c_3 \|x\|^2 \le c_3 V(x) \le \tfrac{\varepsilon}{4} \|\tilde{\theta}\|^2 - \dot{V}. \tag{6.21}$$

Integrating the above over a finite time interval $[0, t]$ yields

$$c_1 c_3 \int_0^t \|x(s)\|^2 ds \le \frac{\varepsilon}{4} \int_0^t \|\tilde{\theta}(s)\|^2 ds - V(x(t)) + V(x(0))$$

$$\le \frac{\varepsilon}{4} \int_0^t \|\tilde{\theta}(s)\|^2 ds + V(x(0)).$$

Taking limits as $t \to \infty$ and noting that $\tilde{\theta}(\cdot) \in \mathcal{L}_2$ by Lemma 6.1 yields

$$\int_0^\infty \|x(s)\|^2 ds \le \frac{\varepsilon}{4 c_1 c_3} \int_0^\infty \|\tilde{\theta}(s)\|^2 ds + \frac{V(x(0))}{c_1 c_3} < \infty,$$

implying $x(\cdot) \in \mathcal{L}_2$. It follows from Lemma 6.1 that $\tilde{\theta}(\cdot) \in \mathcal{L}_\infty$ and thus $\hat{\theta}(\cdot) \in \mathcal{L}_\infty$. Combining this with the assumption that $u = k(x, \hat{\theta})$ is bounded and $x(\cdot) \in \mathcal{L}_\infty$ implies that $\dot{x}(\cdot) \in \mathcal{L}_\infty$. Since $x(\cdot), \dot{x}(\cdot) \in \mathcal{L}_\infty$ and $x(\cdot) \in \mathcal{L}_2$, Lemma 6.2 implies $\lim_{t \to \infty} x(t) = 0$. $\square$

The above theorem only certifies asymptotic stability whereas the exponentially stabilizing adaptive CLF (ES-aCLF) controllers posed in Chap. 4 enforce exponential stability. It should be noted, however, that the exponential stability results of Chap. 4 are established with respect to a *composite* system consisting of the original dynamical system and the parameter estimation error dynamics. The ISS approach presented herein has the benefit of characterizing the transient behavior of the system trajectory rather than a composite system trajectory. Moreover, as discussed earlier in this chapter, taking the ISS approach outlined above has certain benefits when combined with CBF-based adaptive controllers, as demonstrated in the subsequent section.

6.3 Input-to-State Safety

In this section we present an extension of the ISS formalism to safety specifications using the notion of *input-to-state safety* (ISSf). Similar to ISS, the framework of ISSf allows for characterizing the degradation of safety guarantees in the presence of uncertainties. In particular, the ISSf framework is concerned with establishing forward invariance of an *inflated* safe set $\mathcal{C}_\delta \supset \mathcal{C}$ whose inflation is proportional to the magnitude of uncertainty perturbing the nominal system dynamics. Formally, given a continuously differentiable function $h : \mathbb{R}^n \to \mathbb{R}$ we define the inflated safe set for some $\delta \in \mathbb{R}_{\geq 0}$ as

$$
\begin{aligned}
\mathcal{C}_\delta &:= \{x \in \mathbb{R}^n \mid h(x) + \gamma(\delta) \geq 0\}, \\
\partial\mathcal{C}_\delta &:= \{x \in \mathbb{R}^n \mid h(x) + \gamma(\delta) = 0\}, \\
\mathrm{Int}(\mathcal{C}_\delta) &:= \{x \in \mathbb{R}^n \mid h(x) + \gamma(\delta) > 0\},
\end{aligned}
\tag{6.22}
$$

where $\gamma \in \mathcal{K}_\infty$. The fact that $\gamma \in \mathcal{K}_\infty$ implies that $\mathcal{C} = \mathcal{C}_\delta$, with $\mathcal{C}$ as in (3.3), when $\delta = 0$, implying that we recover the original safe set in the absence of any uncertainty. We first define the notion of ISSf for the uncertain dynamical system (6.2) and then extend it to our system of interest (6.1) in the context of control design.

Definition 6.4 (*Input-to-state safety*) Consider system (6.2) and a set $\mathcal{C} \subset \mathbb{R}^n$ as in (3.3). System (6.2) is said to be *input-to-state safe* if there exists $\gamma \in \mathcal{K}_\infty$ and $\delta \in \mathbb{R}_{\geq 0}$ such that for all $d(\cdot)$ satisfying $\|d\|_\infty \leq \delta$, the set $\mathcal{C}_\delta$ in (6.22) is forward invariant.

The above definition provides a pathway towards developing controllers for (6.1) using the notion of an input-to-state safe control barrier function (ISSf-CBF).

Definition 6.5 (*Input-to-state safe CBF*) A continuously differentiable function $h : \mathbb{R}^n \to \mathbb{R}$ defining a set $\mathcal{C} \subset \mathbb{R}^n$ as in (3.3) is said to be an *input-to-state safe control barrier function* for (6.1) on $\mathcal{C}$ if $\nabla h(x) \neq 0$ for all $x \in \partial\mathcal{C}$ and there exists $\alpha \in \mathcal{K}_\infty^e$, $\varepsilon \in \mathbb{R}_{>0}$ such that for all $(x, \hat{\theta}) \in \mathbb{R}^n \times \mathbb{R}^p$

$$
\sup_{u \in \mathcal{U}} \left\{ L_f h(x) + L_F h(x)\hat{\theta} + L_g h(x)u \right\} > -\alpha(h(x)) + \tfrac{1}{\varepsilon}\|L_F h(x)\|^2.
\tag{6.23}
$$

The above definition allows for characterizing the set of all controllers meeting the criteria in (6.23) as

$$
K_{\mathrm{ISSf}}(x, \hat{\theta}) := \left\{ u \in \mathcal{U} \mid L_f h(x) + L_F h(x)\hat{\theta} + L_g h(x)u \geq -\alpha(h(x)) + \tfrac{1}{\varepsilon}\|L_F h(x)\|^2 \right\}.
\tag{6.24}
$$

We show in the following theorem that any locally Lipschitz controller satisfying $k(x, \hat{\theta}) \in K_{\mathrm{ISSf}}(x, \hat{\theta})$ renders the closed-loop system ISSf.

Theorem 6.4 *Let h be an ISSf-CBF for (6.1) on $\mathcal{C}$ and assume that $\tilde{\theta}(\cdot) \in \mathcal{L}_\infty$ such that $\|\tilde{\theta}\|_\infty \leq \delta$ for some $\delta \in \mathbb{R}_{\geq 0}$. Then, any locally Lipschitz controller $u = k(x, \hat{\theta})$ satisfying $k(x, \hat{\theta}) \in K_{\mathrm{ISSf}}(x, \hat{\theta})$ for all $(x, \hat{\theta}) \in \mathbb{R}^n \times \mathbb{R}^p$ renders $\mathcal{C}_\delta$ forward invariant for the closed-loop system with*

$$
\gamma(\delta) := -\alpha^{-1}\left(-\frac{\varepsilon\delta^2}{4}\right).
\tag{6.25}
$$

Proof Taking the Lie derivative of h along the closed-loop system and lower bounding yields

$$
\begin{aligned}
\dot{h} &= L_f h(x) + L_F h(x)\hat{\theta} + L_g h(x)k(x,\hat{\theta}) + L_F h(x)\tilde{\theta} \\
&\geq -\alpha(h(x)) + \tfrac{1}{\varepsilon}\|L_F h(x)\|^2 + L_F h(x)\tilde{\theta} \\
&\geq -\alpha(h(x)) + \tfrac{1}{\varepsilon}\|L_F h(x)\|^2 - \|L_F h(x)\|\|\tilde{\theta}\|_\infty \\
&\geq -\alpha(h(x)) + \tfrac{1}{\varepsilon}\|L_F h(x)\|^2 - \|L_F h(x)\|\delta \\
&\geq -\alpha(h(x)) - \frac{\varepsilon\delta^2}{4},
\end{aligned}
\tag{6.26}
$$

where the first inequality follows from K_{ISSf}, the second from

$$
L_F h(x)\tilde{\theta} \geq -\|L_F h(x)\|\|\tilde{\theta}\| \geq -\|L_F h(x)\|\|\tilde{\theta}\|_\infty,
$$

the third from $\|\tilde{\theta}\|_\infty \leq \delta$, and the fourth from completing squares. Now define

$$
h_\delta(x,\delta) := h(x) + \gamma(\delta),
\tag{6.27}
$$

and note that $\mathcal{C}_\delta$ is the zero-superlevel set of h_δ. Taking the Lie derivative of h_δ along the closed-loop system and lower bounding yields

$$
\dot{h}_\delta = \dot{h} + \dot{\gamma}(\delta) = \dot{h} \geq -\alpha(h(x)) - \frac{\varepsilon\delta^2}{4}.
\tag{6.28}
$$

Hence, to show that $\mathcal{C}_\delta$ is forward invariant, we must show that $\dot{h} \geq 0$ whenever $x \in \partial\mathcal{C}_\delta$. To this end, for $x \in \partial\mathcal{C}_\delta$ we have

$$
h(x) = -\gamma(\delta) = \alpha^{-1}\left(-\frac{\varepsilon\delta^2}{4}\right),
\tag{6.29}
$$

which implies that, for $x \in \partial\mathcal{C}_\delta$,

$$
\alpha(h(x)) + \frac{\varepsilon\delta^2}{4} = 0,
$$

and it follows from (6.28) that

$$
x \in \partial\mathcal{C}_\delta \implies \dot{h}_\delta \geq 0,
$$

which implies the forward invariance of $\mathcal{C}_\delta$ by Corollary 3.1, as desired. □

Given an ISSf-CBF and a nominal adaptive policy $k_0(x,\hat{\theta})$, e.g., an eISS-CLF-QP controller from earlier in this chapter, an ISSf-CBF-based safety filter can be constructed by solving the optimization problem

$$k(x, \hat{\theta}) = \arg\min_{u \in \mathcal{U}} \quad \frac{1}{2} \|u - k_0(x, \hat{\theta})\|^2$$

$$\text{subject to} \quad L_f h(x) + L_F h(x)\hat{\theta} + L_g h(x)u \geq -\alpha(h(x)) + \frac{1}{\varepsilon}\|L_F h(x)\|^2, \tag{6.30}$$

which is a QP when $\mathcal{U} = \mathbb{R}^m$ or $\mathcal{U}$ is a convex polytope. If $\mathcal{U} = \mathbb{R}^m$, the closed-form expression to (6.30) is given by

$$k(x, \hat{\theta}) = \begin{cases} k_0(x, \hat{\theta}), & \text{if } \psi(x, \hat{\theta}) \geq 0 \\ k_0(x, \hat{\theta}) - \frac{\psi(x,\hat{\theta})}{\|L_g h(x)\|^2} L_g h(x)^\top, & \text{if } \psi(x, \hat{\theta}) < 0, \end{cases} \tag{6.31}$$

where

$$\psi(x, \hat{\theta}) := L_f h(x) + L_F V(x)\hat{\theta} + L_g h(x)k_0(x, \hat{\theta}) + \alpha(h(x)) + \frac{1}{\varepsilon}\|L_F h(x)\|^2.$$

A benefit of the above controller is that a single set of parameters are shared between the nominal (typically performance-based policy) and the safety controller, which is in contrast to the methods developed in Chaps. 4 and 5 that require the parameters to be updated in a particular way to guarantee stability and safety. As with the majority of results outlined thus far, when the parameters in (6.1) are matched and $\mathcal{U} = \mathbb{R}^m$, construction of an ISSf-CBF can be done independently of the uncertain parameters.

Proposition 6.2 *Let $h : \mathbb{R}^n \to \mathbb{R}$ be a CBF for $\dot{x} = f(x) + g(x)u$ with $\mathcal{U} = \mathbb{R}^m$. If the parameters in (6.1) are matched, then h is an ISSf-CBF for (6.1) with $\mathcal{U} = \mathbb{R}^m$.*

As discussed in Chap. 3, the construction of a CBF may be challenging when the user-defined constraint set does not coincide with a controlled invariant safe set. In what follows, we partially[2] address this challenge by demonstrating how to extend the high order CBF (HOCBF) methodology from previous chapters to this ISSf setting. Our development parallels that of Sect. 5.3: we first consider a constraint set

$$\mathcal{C}_0 := \{x \in \mathbb{R}^n \mid h(x) \geq 0\},$$

for some *constraint function* $h : \mathbb{R}^n \to \mathbb{R}$ with relative degree r, and assume that Assumption 5.3 holds so that both the control input and uncertain parameters only appear in $h^{(r)}$, the rth derivative of h along system (6.1). We then recursively construct the collection of functions from (3.17)

$$\psi_0(x) = h(x)$$
$$\psi_i(x) = \dot{\psi}_{i-1}(x) + \alpha_i(\psi_{i-1}(x)), \quad \forall i \in \{1, \ldots, r-1\},$$

---

[2] We only partially address this issue as all of our results regarding the construction of controlled invariant sets using the HOCBF methodology leverage the simplifying assumption that no actuation bounds are present.

where each $\alpha_i \in \mathcal{K}_\infty^e$, which is used to construct the candidate safe set

$$\mathcal{C} := \bigcap_{i=0}^{r-1} \mathcal{C}_i,$$

where

$$\mathcal{C}_i := \{x \in \mathbb{R}^n \mid \psi_i(x) \geq 0\}, \quad \forall i \in \{0, \ldots, r-1\}.$$

To extend the HOCBF framework to this ISSf setting, we define a collection of inflated sets

$$\mathcal{C}_{\delta,i} := \{x \in \mathbb{R}^n \mid \psi_i(x) + \gamma_i(\delta) \geq 0\}, \quad \forall i \in \{0, \ldots, r-1\} \tag{6.32}$$

which is used to define the overall inflated safe set

$$\mathcal{C}_\delta := \bigcap_{i=0}^{r-1} \mathcal{C}_{\delta,i}, \tag{6.33}$$

where each $\gamma_i \in \mathcal{K}_\infty$, whose controlled invariance we wish to certify using the notion of an *input-to-state safe high order control barrier function* (ISSf-HOCBF).

Definition 6.6 (*ISSf high order CBF*) Let $h : \mathbb{R}^n \to \mathbb{R}$ have relative degree $r \in \mathbb{N}$ at some $x \in \mathbb{R}^n$ for (6.1) such that $\nabla \psi_i(x) \neq 0$ for all $x \in \partial \mathcal{C}_i$ for each $i \in \{0, \ldots, r-1\}$ and let Assumption 5.3 hold. The function h is said to be an *input-to-state safe high order control barrier function* for (6.1) on $\mathcal{C}$ as in (3.19) if there exist $\alpha_r \in \mathcal{K}_\infty^e$, $\varepsilon \in \mathbb{R}_{>0}$ such that for all $(x, \hat{\theta}) \in \mathbb{R}^n \times \mathbb{R}^p$

$$\sup_{u \in \mathcal{U}} \{L_f \psi_{r-1}(x) + L_F \psi_{r-1}(x)\hat{\theta} + L_g \psi_{r-1}(x)u\}$$
$$> -\alpha_r(\psi_{r-1}(x)) + \frac{1}{\varepsilon} \|L_F \psi_{r-1}(x)\|^2. \tag{6.34}$$

The above definition allows for constructing the set of all control policies satisfying the criteria in (6.34) as

$$K_\psi(x, \hat{\theta}) := \{u \in \mathcal{U} \mid L_f \psi_{r-1}(x) + L_F \psi_{r-1}(x)\hat{\theta} + L_g \psi_{r-1}(x)u$$
$$\geq -\alpha_r(\psi_{r-1}(x)) + \frac{1}{\varepsilon} \|L_F \psi_{r-1}(x)\|^2\}, \tag{6.35}$$

and we show in the following theorem that any locally Lipschitz policy belonging to the above set renders $\mathcal{C}_\delta$ forward invariant.

Theorem 6.5 *Let h be an ISSf-HOCBF for (6.1) on $\mathcal{C} \subset \mathbb{R}^n$ as in (3.19) and assume that $\tilde{\theta}(\cdot)$ is such that $\|\tilde{\theta}\|_\infty \leq \delta$ for some $\delta \in \mathbb{R}_{\geq 0}$. Then, any locally Lipschitz controller $u = k(x, \hat{\theta})$ satisfying $k(x, \hat{\theta}) \in K_\psi(x, \hat{\theta})$ for all $(x, \hat{\theta}) \in \mathbb{R}^n \times \mathbb{R}^p$, with K_ψ as in (6.35), renders $\mathcal{C}_\delta$ from (6.33) forward invariant for the closed-loop system with*

$$\gamma_{r-1}(\delta) := -\alpha_r^{-1}\left(-\frac{\varepsilon\delta^2}{4}\right)$$

(6.36)

$$\gamma_i(\delta) := -\alpha_{i+1}^{-1}(-\gamma_{i+1}(\delta)), \quad \forall i \in \{0, \ldots, r-2\}.$$

Proof The proof is approached in the same manner as that of Theorem 6.4. Taking the Lie derivative of ψ_{r-1} along the closed-loop system and lower bounding yields

$$
\begin{aligned}
\dot{\psi}_{r-1} &= L_f\psi_{r-1}(x) + L_F\psi_{r-1}(x)\hat{\theta} + L_g\psi_{r-1}(x)k(x,\hat{\theta}) + L_F\psi_{r-1}(x)\tilde{\theta}\\
&\geq -\alpha_r(\psi_{r-1}(x)) + \frac{1}{\varepsilon}\|L_F\psi_{r-1}(x)\|^2 + L_F\psi_{r-1}(x)\tilde{\theta}\\
&\geq -\alpha_r(\psi_{r-1}(x)) + \frac{1}{\varepsilon}\|L_F\psi_{r-1}(x)\|^2 - \|L_F\psi_{r-1}(x)\|\|\tilde{\theta}\|_\infty\\
&\geq -\alpha_r(\psi_{r-1}(x)) + \frac{1}{\varepsilon}\|L_F\psi_{r-1}(x)\|^2 - \|L_F\psi_{r-1}(x)\|\delta\\
&\geq -\alpha_r(\psi_{r-1}(x)) - \frac{\varepsilon\delta^2}{4},
\end{aligned}
$$

(6.37)

where the first inequality follows from K_ψ, the second from

$$L_F\psi_{r-1}(x)\tilde{\theta} \geq -\|L_F\psi_{r-1}(x)\|\|\tilde{\theta}\| \geq -\|L_F\psi_{r-1}(x)\|\|\tilde{\theta}\|_\infty,$$

the third from $\|\tilde{\theta}\|_\infty \leq \delta$, and the fourth from completing squares. Now define

$$\psi_{\delta,r-1}(x,\delta) := \psi_{r-1}(x) + \gamma_{r-1}(\delta),$$

(6.38)

and note that $\mathcal{C}_{\delta,r-1}$ is the zero-superlevel set of $\psi_{\delta,r-1}$. Taking the Lie derivative of $\psi_{\delta,r-1}$ along the closed-loop system and lower bounding yields

$$\dot{\psi}_{\delta,r-1} = \dot{\psi}_{r-1} + \dot{\gamma}_{r-1}(\delta) = \dot{\psi}_{r-1} \geq -\alpha_r(\psi_{r-1}(x)) - \frac{\varepsilon\delta^2}{4}.$$

(6.39)

To show that $\mathcal{C}_{\delta,r-1}$ is forward invariant we must show that $x \in \partial\mathcal{C}_{\delta,r-1} \implies \dot{\psi}_{r-1} \geq 0$. To this end, observe that

$$x \in \partial\mathcal{C}_{\delta,r-1} \implies \psi_{r-1}(x) = -\gamma_{r-1}(\delta),$$

so that

$$x \in \partial\mathcal{C}_{\delta,r-1} \implies \dot{\psi}_{r-1} \geq -\alpha_r(-\gamma_{r-1}(\delta)) - \frac{\varepsilon\delta^2}{4}.$$

Taking γ_{r-1} as in (6.36), we have

$$x \in \partial\mathcal{C}_{\delta,r-1} \implies \dot{\psi}_{r-1} \geq -\alpha_r\left(\alpha_r^{-1}\left(-\frac{\varepsilon\delta^2}{4}\right)\right) - \frac{\varepsilon\delta^2}{4} = 0.$$

It then follows from Corollary 3.1 that $\mathcal{C}_{\delta,r-1}$ is forward invariant for the closed-loop system, which implies that the closed-loop trajectory $t \mapsto x(t)$ satisfies

$$\psi_{r-1}(x(t)) \geq -\gamma_{r-1}(\delta), \quad \forall t \in I(x(0)),$$

where $I(x(0)) \subset \mathbb{R}_{\geq 0}$ is the trajectory's maximal interval of existence from an initial condition of $x(0) \in \mathcal{C}_\delta$. Using the definition of ψ_{r-1} from (3.17), we then have that

$$\psi_{r-1}(x(t)) = \dot{\psi}_{r-2}(x(t), \tilde{\theta}(t)) + \alpha_{r-1}(\psi_{r-2}(x(t))) \geq -\gamma_{r-1}(\delta), \quad \forall t \in I(x(0)),$$

which implies that

$$\dot{\psi}_{r-2} \geq -\alpha_{r-1}(\psi_{r-2}(x(t))) - \gamma_{r-1}(\delta), \quad \forall t \in I(x(0)).$$

Now define

$$\psi_{\delta,r-2}(x, \delta) := \psi_{r-2}(x) + \gamma_{r-2}(\delta), \tag{6.40}$$

and note that $\mathcal{C}_{\delta,r-2}$ is the zero-superlevel set of $\psi_{\delta,r-2}$. Taking the Lie derivative of $\psi_{\delta,r-2}$ along the closed-loop system and lower bounding yields

$$\dot{\psi}_{\delta,r-2} = \dot{\psi}_{r-2} + \dot{\gamma}_{r-2}(\delta) = \dot{\psi}_{r-2} \geq -\alpha_{r-1}(\psi_{r-2}(x(t))) - \gamma_{r-1}(\delta). \tag{6.41}$$

Hence, to show that $\mathcal{C}_{\delta,r-2}$ is forward invariant we must show that $x \in \partial\mathcal{C}_{\delta,r-2} \implies \dot{\psi}_{r-2} \geq 0$. To this end, observe that

$$x \in \partial\mathcal{C}_{\delta,r-2} \implies \psi_{r-2}(x) = -\gamma_{r-2}(\delta),$$

so that

$$x \in \partial\mathcal{C}_{\delta,r-2} \implies \dot{\psi}_{r-2} \geq -\alpha_{r-1}(-\gamma_{r-2}(\delta)) - \gamma_{r-1}(\delta).$$

Taking γ_{r-2} as in (6.36), we have

$$x \in \partial\mathcal{C}_{\delta,r-2} \implies \dot{\psi}_{r-2} \geq -\alpha_{r-1}\left(\alpha_{r-1}^{-1}(-\gamma_{r-1}(\delta))\right) - \gamma_{r-1}(\delta) = 0.$$

It then follows from Corollary 3.1 that $\mathcal{C}_{\delta,r-2}$ is forward invariant for the closed-loop system. One can then take analogous steps to those outlined above for the remaining ψ_i terms to show that, provided $x(0) \in \mathcal{C}_\delta$, then $\psi_{\delta,i}(x(t)) \geq 0$ for all $t \in I(x(0))$ and all $i \in \{0, \ldots, r-1\}$, which implies $x(t) \in \mathcal{C}_\delta$ for all $t \in I(x(0))$, as desired. $\square$

As with ISSf-CBFs, given an ISSf-HOCBF and a nominal adaptive policy $k_0(x, \hat{\theta})$, a safety filter enforcing the invariance of $\mathcal{C}_\delta$ can be constructed by solving the optimization problem

$$\begin{aligned}
k(x, \hat{\theta}) = \arg\min_{u \in \mathcal{U}} \quad & \frac{1}{2}\|u - k_0(x, \hat{\theta})\|^2 \\
\text{subject to} \quad & L_f\psi_{r-1}(x) + L_F\psi_{r-1}(x)\hat{\theta} + L_g\psi_{r-1}(x)u \\
& \geq -\alpha_r(\psi_{r-1}(x)) + \frac{1}{\varepsilon}\|L_F\psi_{r-1}(x)\|^2,
\end{aligned} \tag{6.42}$$

which is again a QP when $\mathcal{U} = \mathbb{R}^m$ or $\mathcal{U}$ is a convex polytope. Moreover, when $\mathcal{U} = \mathbb{R}^m$, the above QP has a closed-form solution, which is given by

$$k(x, \hat{\theta}) = \begin{cases} k_0(x, \hat{\theta}), & \text{if } \Psi(x, \hat{\theta}) \geq 0 \\ k_0(x, \hat{\theta}) - \frac{\Psi(x,\hat{\theta})}{\|L_g\psi_{r-1}(x)\|^2} L_g\psi_{r-1}(x)^\top, & \text{if } \Psi(x, \hat{\theta}) < 0, \end{cases} \tag{6.43}$$

where

$$\Psi(x, \hat{\theta}) := L_f\psi_{r-1}(x) + L_F\psi_{r-1}(x)\hat{\theta} + L_g\psi_{r-1}(x)k_0(x, \hat{\theta})$$
$$+ \alpha_r(\psi_{r-1}(x)) - \tfrac{1}{\varepsilon}\|L_F\psi_{r-1}(x)\|^2.$$

Similar to ISSf-CBFs, when the parameters in (6.1) are matched and $\mathcal{U} = \mathbb{R}^m$, an ISSf-HOCBF can be constructed using a HOCBF for the nominal system dynamics:

Proposition 6.3 *Let* $h : \mathbb{R}^n \to \mathbb{R}$ *be a HOCBF for* $\dot{x} = f(x) + g(x)u$ *with* $\mathcal{U} = \mathbb{R}^m$. *If the parameters in* (6.1) *are matched, then* h *is a ISSf-HOCBF for* (6.1).

Recall that care must be taken when constructing HOCBFs as even seemingly benign safety constraints may produce invalid HOCBFs without further modifications (see Sect. 3.3).

6.4   Numerical Examples

Example 6.1 We illustrate the methods developed in this chapter using a simple obstacle avoidance scenario for a planar mobile robot modeled as a double integrator with nonlinear drag effects of the form

$$\ddot{q} = -D\dot{q}\|\dot{q}\| + u, \tag{6.44}$$

where $q \in \mathbb{R}^2$ denotes the robot's position, $u \in \mathbb{R}^2$ its commanded acceleration, and $D \in \mathbb{R}^{2\times2}$ a diagonal matrix of damping coefficients. Defining $x := [q^\top \ \dot{q}^\top]^\top \in \mathbb{R}^4$ allows (6.44) to be represented in the form of (4.1) as

$$\dot{x} = \underbrace{\begin{bmatrix} \dot{q} \\ 0 \end{bmatrix}}_{f(x)} + \underbrace{\begin{bmatrix} 0_{2\times2} \\ \text{diag}(\dot{q}\|\dot{q}\|) \end{bmatrix}}_{F(x)} \underbrace{\begin{bmatrix} D_1 \\ D_2 \end{bmatrix}}_{\theta} + \underbrace{\begin{bmatrix} 0_{2\times2} \\ I_{2\times2} \end{bmatrix}}_{g(x)} u, \tag{6.45}$$

where $0_{2\times2} \in \mathbb{R}^{2\times2}$ is a 2×2 matrix of zeros, $I_{2\times2}$ is a 2×2 identity matrix, $\text{diag}(\cdot)$ constructs a diagonal matrix from a vector, and $D_1, D_2 \in \mathbb{R}_{>0}$ are the unknown drag coefficients. Our control objective is to drive (6.44) to the origin while avoiding an obstacle in the workspace and learning the uncertain parameters online. To estimate the uncertain parameters online, we leverage the concurrent learning approach introduced in Chap. 4. Recall that such an approach is predicated on the observation that, along state-control trajectory $(x(\cdot), u(\cdot))$, system (4.1) can be expressed as

$$\int_{\max\{t-\Delta t,0\}}^{t} \dot{x}(s)ds = \int_{\max\{t-\Delta t,0\}}^{t} f(x(s))ds + \int_{\max\{t-\Delta t,0\}}^{t} F(x(s))ds\theta$$

$$+ \int_{\max\{t-\Delta t,0\}}^{t} g(x(s))u(s)ds,$$

for all $t \geq 0$, where $\Delta t \in \mathbb{R}_{>0}$ is the length of an integration window. Defining

$$\mathcal{Y}(t) := \int_{\max\{t-\Delta t,0\}}^{t} (\dot{x}(s) - f(x(s)) - g(x(s))u(s))ds,$$

$$\mathcal{F}(t) := \int_{\max\{t-\Delta t,0\}}^{t} F(x(s))ds$$

yields the linear relationship for the uncertain parameters

$$\mathcal{Y}(t) = \mathcal{F}(t)\theta. \tag{6.46}$$

The parameters can then be recursively estimated online by storing values of $\mathcal{Y}$ and $\mathcal{F}$ at run-time in a history stack $\mathcal{H} = \{(\mathcal{Y}_j, \mathcal{F}_j)\}_{j=1}^{M}$ using Algorithm 1 from Chap. 4, which can then be used in a parameter update law $\dot{\hat{\theta}} = \tau(\hat{\theta}, t)$ to improve the parameter estimates. For example, in previous chapters such update laws were derived by forming the prediction error

$$e(\hat{\theta}) = \sum_{j=1}^{M} \|\mathcal{Y}_j - \mathcal{F}_j\hat{\theta}\|^2,$$

based on the data available in $\mathcal{H}$ and then recursively minimizing such an error online using gradient descent. Rather than taking such an approach, we demonstrate how any estimation algorithm satisfying the conditions of Lemma 6.1 can be used to endow the resulting adaptive control system with stability and safety guarantees. To this end, we consider the following class of update laws

$$\dot{\hat{\theta}} = -\Gamma(t)\nabla e(\hat{\theta}) = \Gamma(t) \sum_{j=1}^{N} \mathcal{F}_j^{\top}(\mathcal{Y}_j - \mathcal{F}_j\hat{\theta}), \tag{6.47}$$

which serves as a general template for particular update laws based on the properties of $\Gamma(\cdot)$ as follows:

$$\dot{\Gamma} = 0, \tag{6.48a}$$

$$\dot{\Gamma} = -\Gamma\left[\sum_{j=1}^{N} \mathcal{F}_j^{\top}\mathcal{F}_j\right]\Gamma, \tag{6.48b}$$

$$\dot{\Gamma} = \beta\Gamma - \Gamma\left[\sum_{j=1}^{N} \mathcal{F}_j^{\top}\mathcal{F}_j\right]\Gamma, \tag{6.48c}$$

$$\dot{\Gamma} = \beta \left[1 - \frac{\|\Gamma\|}{\bar{\Gamma}} \right] \Gamma - \Gamma \left[\sum_{j=1}^{N} \mathcal{F}_j^\top \mathcal{F}_j \right] \Gamma, \qquad (6.48d)$$

where $\beta \in \mathbb{R}_{>0}$ is a forgetting/discount factor and $\bar{\Gamma} \in \mathbb{R}_{>0}$ is a user-defined constant that bounds $\|\Gamma(t)\|$. With $\dot{\Gamma}$ as in (6.48), the update law in (6.47) corresponds to: (6.48a) gradient descent; (6.48b) recursive least squares (RLS); (6.48c) RLS with a forgetting/discount factor; (6.48d) RLS with a variable forgetting factor. We emphasize that the purpose of our numerical example is not necessarily to establish superiority of one algorithm over the others; rather, our goal is to demonstrate that, under the assumptions posed in Lemma 6.1, the stability/safety guarantees of the controller can be decoupled from the design of the parameter estimator, which allows considerable freedom in selecting an estimation algorithm best suited for the problem at hand. We demonstrate the modularity of our approach (i.e., the ability to decouple the design of the estimator from the controller) by running a set of simulations with randomly sampled initial conditions for the system state and estimated parameters under each algorithm, and show that, for a given level of uncertainty, the ISSf guarantees are invariant to the particular choice of parameter estimator. For each estimation algorithm we produce 25 different trajectories by uniformly sampling the initial state from $[-1.8, -2.2] \times [1.8, 2.2] \times \{0\} \times \{0\} \subset \mathbb{R}^4$ and the initial parameter estimates from $[0, 3]^2 \subset \mathbb{R}^2$; the true parameters are set to $\theta = [0.8\ 1.4]^\top$. The hyperparameters for the estimation algorithms are selected as $N = 20$, $\Gamma(0) = 100 I_{2 \times 2}$, $\beta = 1$, $\bar{\Gamma} = 1000$. The stabilization objective is achieved by considering the eISS-CLF candidate $V(x) = \frac{1}{2}\|q\|^2 + \frac{1}{2}\|q + \dot{q}\|^2$ with $c_3 = 1$ and $\varepsilon = 20$. The safety objective is achieved by considering the constraint function $h(x) = \|q - q_o\|^2 - R_o^2$, where $q_o = [-1\ 1]^\top$ is the center of the circular obstacle and $R_o = 0.5$ its radius, which has relative degree 2 for (6.44) with respect to both u and θ. This constraint function is used to construct an ISSf-HOCBF candidate with $\alpha_1(s) = s$, $\alpha_2(s) = \frac{1}{2}s$, and $\varepsilon = 1$. To determine the validity of h as a valid ISSf-HOCBF we note that as the parameters in (6.44) are matched, it suffices to show that h is a valid HOCBF for a two-dimensional double integrator without any uncertainty. To this end, we compute Lie derivatives of h along the system dynamics

$$\nabla h(x) = \begin{bmatrix} 2(q - q_0) \\ 0_{2 \times 1} \end{bmatrix}$$

$$L_f h(x) = \begin{bmatrix} 2(q - q_0)^\top & 0_{1 \times 2} \end{bmatrix} \begin{bmatrix} \dot{q} \\ 0_{2 \times 1} \end{bmatrix} = 2(q - q_0)^\top \dot{q}$$

$$L_g h(x) = \begin{bmatrix} 2(q - q_0)^\top & 0_{1 \times 2} \end{bmatrix} \begin{bmatrix} 0_{2 \times 2} \\ I_{2 \times 2} \end{bmatrix} = 0_{1 \times 2},$$

and see that h has relative degree larger than one. Computing second order Lie derivatives yields

$$\nabla L_f h(x) = \begin{bmatrix} 2\dot{q} \\ 2(q - q_0) \end{bmatrix}$$

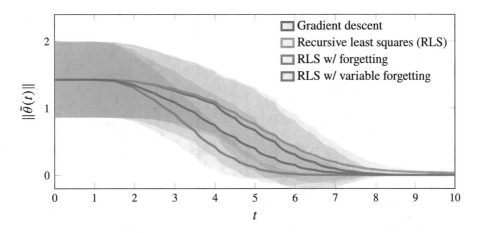

Fig. 6.1 Mean and standard deviation of the norm of the parameter estimation over time generated by each parameter estimator. The solid lines indicate the average value of $\|\tilde{\theta}(t)\|$ across each simulation, and the ribbon surrounding each line corresponds to one standard deviation from the mean

$$L_g L_f h(x) = \begin{bmatrix} 2\dot{q}^\top & 2(q - q_0)^\top \end{bmatrix} \begin{bmatrix} 0_{2\times2} \\ I_{2\times2} \end{bmatrix} = 2(q - q_0)^\top,$$

which reveals that $L_g L_f h(x) \neq 0$ for all states whose position does not lie at the center of the obstacle. Hence, h has relative degree 2 for all $x \in \mathcal{C}$, where $\mathcal{C}$ is defined recursively by h as in (3.19), and h is thus a valid ISSf-HOCBF.

For each simulation, the closed-loop trajectory is generated by the ISSf-HOCBF controller in (6.42), where the nominal adaptive policy is chosen as the eISS-CLF controller from (6.12), the results of which are provided in Figs. 6.1 and 6.2. As shown in Fig. 6.2, the trajectories under each update law remain safe and converge to the origin, whereas Fig. 6.1 illustrates the convergence of the parameter estimation error to zero for each estimation algorithm as predicted by Lemma 6.1. The curves in Fig. 6.1 represent the mean and standard deviation of the parameter estimation error over time across all simulations for each estimation algorithm. The results in Fig. 6.1 illustrate that, on average, the RLS with forgetting factor estimator (6.48c) produces the fastest convergence of the parameters estimates while also exhibiting low variance across different trajectories. The standard RLS algorithm (6.48b) produces the slowest convergence, which is expected given that, in general, this algorithm cannot guarantee exponential convergence[3] of the parameter estimates, whereas the others can.

---

[3] This also implies that (6.48b) does not satisfy all the conditions of Lemma 6.1. Despite this, note that boundedness of the estimates is sufficient to establish ISS and ISSf.

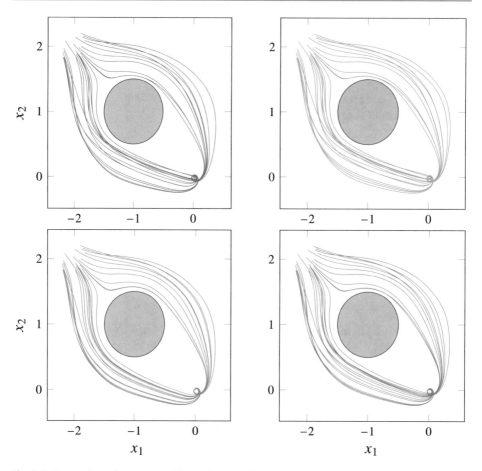

Fig. 6.2 State trajectories generated by each estimation algorithm projected onto the x_1-x_2 plane. In each plot the gray disk denotes the obstacle. The colors in each plot share the same interpretation as those in Fig. 6.1

Example 6.2 In the preceding examples, safety was enforced by choosing an appropriate value of ε for the given level of uncertainty. In theory, $\mathcal{C}_\delta \to \mathcal{C}$ as $\varepsilon \to 0$; however, taking ε very small may require a significant amount of control effort that could exceed physical actuator limits. An alternative approach to reducing safety violations in this ISSf setting is through fast adaptation - if the parameter estimates quickly converge to their true values then the estimated dynamics used in (6.42) to generate control actions will be very close to the true dynamics. In Fig. 6.3, we generated additional trajectories of the closed-loop system under the gradient descent update law (6.48a) and the RLS update law with a forgetting factor (6.48c) using the same setup as in the previous example, but with different levels of initial parameter uncertainty. As demonstrated in Fig. 6.3, the trajectories under the RLS update law avoid the obstacle for the given initial parameter estimation errors via fast adaptation,

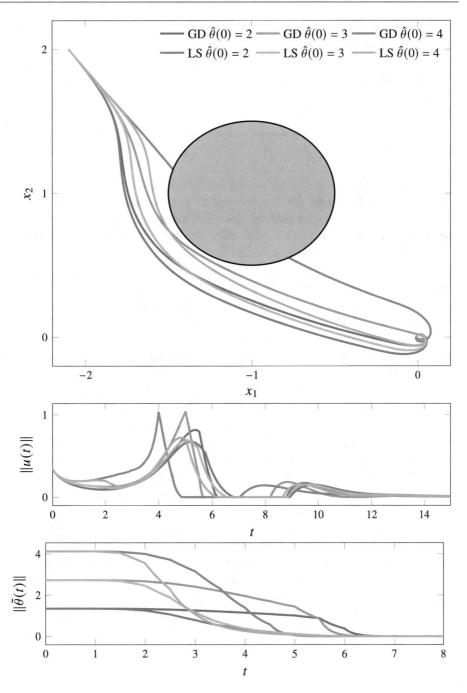

Fig. 6.3 Comparison between trajectories generated by the gradient descent (GD) learning algorithm (6.48a) and the recursive least squares algorithm (LS) with a forgetting factor (6.48c) for different initial parameter estimates. The top plot displays the system trajectories, the middle illustrates the control trajectories, and the bottom illustrates the parameter estimation error

whereas the trajectories under the gradient descent algorithm violate the safety constraint for higher levels of uncertainty. Hence, rather than using a more robust controller (by decreasing ε), which may be overly conservative if bounds on θ are unknown, one can endow the ISSf controller with stronger safety guarantees through the use of a more efficient estimation algorithm.

6.5   Notes

In this chapter we introduced a class of adaptive control techniques that are referred to as *modular* in the sense that the design of the controller can be decoupled from the design of the parameter estimator. Such a property allows for interchanging the parameter estimation algorithm with minimal impact on the resulting stability/safety guarantees. Whereas the adaptive control methods from Chap. 4 could be classified as Lyapunov-based, the approach presented in the present chapter can generally be classified as an estimation-based approach since the primary objective of the parameter estimator is often to reduce the prediction error.

The modular approach to adaptive stabilization in this chapter relies heavily on the notion of *input-to-state stability* (ISS), a concept introduced by Sontag in [1]. Since its inception in 1989, ISS has proven to be a powerful tool in nonlinear systems and control - a more complete exposition on ISS and its applications can be found in [2]. Over the years, various notions of ISS control Lyapunov functions (ISS-CLF) have been introduced, see, e.g., [3, 4] for examples. Our definition of an exponential ISS-CLF is inspired by that from [5], where such a function was used to control bipedal robots in the presence of uncertainty. Similar to the CLFs from Chap. 2, it was shown in [3] that ISS-CLFs are also inverse optimal [6] in that they solve a differential game with a meaningful cost function. The idea of using ISS-CLFs for adaptive nonlinear stabilization can be traced back to [7, 8], in which they were combined with general classes of parameter estimators to achieve ISS of nonlinear systems with respect to the parameter estimation error. Extending such modular designs to concurrent learning-based parameter estimators [9] was explored in [10].

Efforts towards extending to ISS paradigm from stability to safety was first explored in [11], where it was shown that, in the presence of disturbances, the stability guarantees induced by CBFs can be given an ISS characterization. These ideas were formalized using the concept of input-to-state safety (ISSf) in [12], which also introduced the notion of an ISSf-CBF. Generalizations of the original ISSf framework along with different notions of ISSf-CBFs have appeared in [13–15]. The idea of using ISSf to extend traditional modular adaptive control approaches to a safety-critical setting first appeared in [10]. Beyond the integration of ISSf and adaptive control, the ISSf framework has also shown promise in other learning-based control frameworks [16–18] as well as in event-triggered control [19].

References

1. Sontag ED (1989) Smooth stabilization implies coprime factorization. IEEE Trans Autom Control 34(4):435–443
2. Sontag ED (2008) Input to state stability: basic concepts and results. In: Nistri P, Stefani G (eds) Nonlinear and optimal control theory. Springer, pp 163–220
3. Krstic M, Li ZH (1998) Inverse optimal design of input-to-state stabilizing nonlinear controllers. IEEE Trans Autom Control 43(3):336–350
4. Liberzon D, Sontag ED, Wang Y (2002) Universal construction of feedback laws achieving iss and integral-iss disturbance attenuation. Syst Control Lett 46:111–127
5. Kolathaya S, Reher J, Hereid A, Ames AD (2018) Input to state stabilizing control lyapunov functions for robust bipedal robotic locomotion. In: Proceedings of the American control conference
6. Freeman RA, Kokotovic PV (1996) Inverse optimality in robust stabilization. SIAM J Control Optim 34(4):1365–1391
7. Krstić M, Kokotović P (1995) Adaptive nonlinear design with controller-identification separation and swapping. IEEE Trans Autom Control 40(3):426–440
8. Krstić M, Kokotović P (1996) Modular approach to adaptive nonlinear stabilization. Automatica 32(4):625–629
9. Chowdhary G (2010) Concurrent learning for convergence in adaptive control without persistency of excitation. PhD thesis, Georgia Institute of Technology, Atlanta, GA
10. Cohen MH, Belta C (2023) Modular adaptive safety-critical control. In: Proceedings of the American control conference
11. Xu X, Tabuada P, Grizzle JW, Ames AD (2015) Robustness of control barrier functions for safety critical control. In: Proceedings of the IFAC conference on analysis and design of hybrid systems, pp 54–61
12. Kolathaya S, Ames AD (2019) Input-to-state safety with control barrier functions. IEEE Control Syst Lett 3(1):108–113
13. Taylor AJ, Singletary A, Yue Y, Ames AD (2020) A control barrier perspective on episodic learning via projection-to-state safety. IEEE Control Syst Lett 5(3):1019–1024
14. Alan A, Taylor AJ, He CR, Orosz G, Ames AD (2022) Safe controller synthesis with tunable input-to-state safe control barrier functions. IEEE Control Syst Lett 6:908–913
15. Alan A, Taylor AJ, He CR, Ames AD, Orosz G (2022) Control barrier functions and input-to-state safety with application to automated vehicles. arXiv:2206.03568
16. Csomay-Shanklin N, Cosner RK, Dai M, Taylor AJ, Ames AD (2021) Episodic learning for safe bipedal locomotion with control barrier functions and projection-to-state safety. In: Proceedings of the 3rd annual conference on learning for dynamics and control, vol 144. Proceedings of machine learning research, pp 1041–1053
17. Cosner RK, Tucker M, Taylor AJ, Li K, Molnar TG, Ubellacker W, Alan A, Orosz G, Yue Y, Ames AD (2022) Safety-aware preference-based learning for safety-critical control. In: 4th annual conference on learning for dynamics and control, vol 166. Proceedings of machine learning research, pp 1–14

18. Cosner RK, Yue Y, Ames AD (2022) End-to-end imitation learning with safety guarantees using control barrier functions. In: Proceedings of the ieee conference on decision and control, pp 5316–5322
19. Taylor AJ, Ong P, Cortés J, Ames AD (2021) Safety-critical event triggered control via input-to-state safe barrier functions. IEEE Control Syst Lett 5(3):749–754

Robust Safety-Critical Control for Systems with Actuation Uncertainty

In earlier chapters, our main objective was to design controllers for nonlinear systems with parametric uncertainty so that the closed-loop system satisfied desired performance requirements such as exponential stability or safety. We focused on the case when the parameters entered the dynamics in an *additive* fashion, in the sense that the dynamics were affine in both the control input and the uncertain parameters. In this chapter, we consider the case when uncertainty enters the dynamics *multiplicatively*, which is relevant to many application areas. We propose a duality-based approach to robust safety-critical control in Sect. 7.1, which is based on robust control barrier functions (Sect. 7.1.1) and robust control Lyapunov functions (Sect. 7.1.2). An online learning approach for uncertainty reduction based on leveraging input-output data generated by the system at run-time is presented in Sect. 7.2. Numerical examples are included in Sect. 7.3. We conclude with final remarks, references, and suggestions for further reading in Sect. 7.4.

In previous chapters, we considered uncertain nonlinear systems of the form (4.1):

$$\dot{x} = f(x) + F(x)\theta + g(x)u,$$

where $\theta \in \mathbb{R}^p$ are constant, but unknown parameters of the underlying dynamical system. In the above system, the parameters enter the dynamics in an *additive* fashion in the sense that the dynamics are affine in both the control input and uncertain parameters. A more challenging situation arises when the uncertainty enters the dynamics *multiplicatively* in the sense that the dynamics are bilinear in the control and parameters:

$$\dot{x} = f(x) + g(x)u + \varphi(x, u)\theta, \tag{7.1}$$

where $\varphi : \mathbb{R}^n \times \mathbb{R}^m \to \mathbb{R}^{n \times p}$ is a locally Lipschitz mapping that is affine in u and $\theta \in \mathbb{R}^p$ once again denotes the uncertain parameters. Allowing the uncertainty to enter the dynamics multiplicatively as in (7.1) is crucial towards extending the ideas introduced thus far to larger

© The Author(s), under exclusive license to Springer Nature Switzerland AG 2023   117
M. Cohen and C. Belta, *Adaptive and Learning-Based Control of Safety-Critical Systems*,
Synthesis Lectures on Computer Science,
https://doi.org/10.1007/978-3-031-29310-8_7

classes of uncertain systems. Even the parameters of relatively simple systems may fail to obey the additive restriction imposed by (4.1).

Example 7.1 Let $q \in \mathbb{R}^3$ be the position of a particle with mass $m \in \mathbb{R}_{>0}$ moving in $\mathbb{R}^3$ acted upon by a control input $u \in \mathbb{R}^3$ whose equations of motion can be derived using Newton's Second Law:

$$m\ddot{q} = u$$

If the mass is unknown, then the above system cannot be put into the form of (4.1) since the uncertain parameters $\theta = \frac{1}{m}$ will multiply the control input. This simple system does, however, fit into the model proposed in (7.1) with state $x = [q^\top \ \dot{q}^\top]^\top$ as

$$\dot{x} = \underbrace{\begin{bmatrix} \dot{q} \\ 0 \end{bmatrix}}_{f(x)} + \underbrace{\begin{bmatrix} 0 \\ u \end{bmatrix}}_{\varphi(x,u)} \underbrace{\frac{1}{m}}_{\theta} .$$

Despite its relevance, system (7.1) presents challenges in designing controllers based on control Lyapunov functions (CLFs) or control barrier functions (CBFs) that are robust to the uncertain parameters. These challenges arise from enforcing the CLF or CBF conditions for all possible realizations of the uncertain parameters, which, as we will demonstrate shortly, does not have the quadratic programming (QP) structure typically used to synthesize such controllers.

7.1 A Duality-Based Approach to Robust Safety-Critical Control

To make the aforementioned challenges more precise, we first impose the following assumption on the parameters:

Assumption 7.1 There exist known constants $\underline{\theta}_i$, $\overline{\theta}_i \in \mathbb{R}$ for all $i \in \{1, \ldots, p\}$ and a hyper-rectangle $\Theta := [\underline{\theta}_1, \overline{\theta}_1] \times \cdots \times [\underline{\theta}_p, \overline{\theta}_p] \subset \mathbb{R}^p$ such that $\theta \in \Theta$.

Assumption 7.1 implies the set of possible parameters Θ admits a halfspace representation as $\Theta = \{\theta \in \mathbb{R}^p \mid A\theta \le b\}$, where A, b capture linear halfspace constraints. As argued in Chap. 5, such an assumption is not restrictive from a practical standpoint as it simply states there exist known bounds on physical attributes of the system such as its inertia and damping properties. In the proceeding sections, we detail the challenges that arise when directly applying CLF/CBF controllers to (7.1) and how such challenges can be overcome using ideas that exploit the duality of a particular class of convex optimization problems.

7.1.1   Robust Control Barrier Functions

In this section, we develop a CBF approach that robustly accounts for all possible realizations of the system uncertainty to design controllers guaranteeing safety of (7.1). Importantly, we show how this can be accomplished while retaining the traditional QP structure used in CBF approaches by exploiting the dual of a particular linear program (LP). As is typically the case when using CBFs, we consider a candidate safe set defined as the zero superlevel set of a continuously differentiable function $h : \mathbb{R}^n \to \mathbb{R}$ as in (3.3):

$$C = \{x \in \mathbb{R}^n \mid h(x) \geq 0\}.$$

We begin by introducing the notion of a robust CBF (RCBF) for systems of the form (7.1).

Definition 7.1 (*Robust CBF*) A continuously differentiable function $h : \mathbb{R}^n \to \mathbb{R}$ is said to be a robust CBF (RCBF) for (7.1) on a set $C \subset \mathbb{R}^n$ as in (3.3) if there exists $\alpha \in \mathcal{K}^e_\infty$ such that for all $x \in \mathbb{R}^n$

$$\sup_{u \in \mathcal{U}} \inf_{\theta \in \Theta} \dot{h}(x, u, \theta) > -\alpha(h(x)), \tag{7.2}$$

where $\dot{h}(x, u, \theta) = L_f h(x) + L_g h(x) u + L_\varphi h(x, u)\theta$.

The above definition states that h is a RCBF for (7.1) if it is possible to enforce the standard CBF condition $\dot{h} \geq -\alpha(h)$ for the worst-case scenario given the feasible parameter set Θ. Similar to the standard CBF case, let

$$K_{\text{rcbf}}(x) := \{u \in \mathcal{U} \mid L_f h(x) + L_g h(x) u + \inf_{\theta \in \Theta} L_\varphi h(x, u)\theta \geq -\alpha(h(x))\}$$

be, for each $x \in \mathbb{R}^n$, the set of control values satisfying the condition from (7.2). The following lemma shows that any locally Lipschitz control policy $k(x) \in K_{\text{rcbf}}(x)$ renders C forward invariant for the closed-loop system.

Lemma 7.1 *If h is a RCBF for (7.1) on a set C as in (3.3) and Assumption 7.1 holds, then any locally Lipschitz control policy $u = k(x)$ satisfying $k(x) \in K_{\text{rcbf}}(x)$ for each $x \in \mathbb{R}^n$ renders C forward invariant for the closed-loop system.*

Proof The derivative of h along the closed-loop system is lower bounded as

$$\begin{aligned}
\dot{h}(x) &= L_f h(x) + L_g h(x) k(x) + L_\varphi h(x, k(x))\theta \\
&\geq L_f h(x) + L_g h(x) k(x) + \inf_{\theta \in \Theta} L_\varphi h(x, k(x))\theta \\
&\geq -\alpha(h(x)).
\end{aligned}$$

Hence, h is a barrier function for the closed-loop system and the forward invariance of $\mathcal{C}$ follows from Theorem 3.2.   □

Although the above lemma demonstrates that the class of CBF from Definition 7.1 provides sufficient conditions for safety, this formulation is not appealing from a control synthesis perspective. In particular, the minimax nature and coupling of control and parameters in Definition 7.1 will lead to bilinear constraints on the control and parameters and thus cannot be directly cast as a QP. For example, directly embedding the conditions imposed by Definition 7.1 into an optimization problem yields

$$\min_{u \in \mathcal{U}} \ \tfrac{1}{2}\|u - k_0(x)\|^2$$
$$\text{subject to} \ \ L_f h(x) + L_g h(x)u + \inf_{\theta \in \Theta} L_\varphi h(x, u)\theta \geq -\alpha(h(x)). \tag{7.3}$$

This optimization problem requires solving simultaneously for both u and θ; however, the constraint is bilinear in u and θ and is thus not a QP. To remedy this, note that the inner minimization problem from (7.2) can be written as the LP[1]:

$$\inf_{\theta} \ \ L_\varphi h(x, u)\theta$$
$$\text{subject to} \ \ A\theta \leq b. \tag{7.4}$$

The dual of (7.4) is

$$\sup_{\mu \leq 0} \ \ b^\top \mu$$
$$\text{subject to} \ \ \mu^\top A = L_\varphi h(x, u), \tag{7.5}$$

where μ is the dual variable. In light of (7.4) and (7.5) we show in Theorem 7.1 that one can solve the following QP

$$\min_{u \in \mathcal{U}, \, \mu \leq 0} \ \ \tfrac{1}{2}\|u - k_d(x)\|^2$$
$$\text{subject to} \ \ L_f h(x) + L_g h(x)u + b^\top \mu \geq -\alpha(h(x)) \tag{7.6}$$
$$\mu^\top A = L_\varphi h(x, u),$$

with decision variables u and μ, to compute a controller satisfying the RCBF conditions from Definition 7.1.

Theorem 7.1 *Let the assumptions of Lemma 7.1 hold. Then any locally Lipschitz solution to (7.6), $u = k(x)$, renders $\mathcal{C}$ forward invariant for the closed-loop system.*

---

[1] Note that $L_\varphi h(x, u)$ is an affine function of u.

Proof The RCBF condition (7.2) is satisfied at a state $x \in \mathbb{R}^n$ if the value of the optimization problem

$$\sup_{u \in \mathcal{U}} \inf_{\theta \in \mathbb{R}^p} \quad L_f h(x) + L_g h(x) u + L_\varphi h(x, u)\theta + \alpha(h(x)) \tag{7.7}$$

$$\text{subject to} \quad A\theta \leq b,$$

is greater than or equal to 0. It follows from the strong duality theorem of LPs that the values of the primal and dual LPs in (7.4) and (7.5), respectively, are equal, allowing the inner minimization in (7.7) to be replaced with its dual (7.5) yielding

$$\sup_{u \in \mathcal{U}, \mu \in \mathbb{R}^{2p}} \quad L_f h(x) + L_g h(x) u + b^\top \mu + \alpha(h(x)) \tag{7.8}$$

$$\text{subject to} \quad \mu^\top A = L_\varphi h(x, u), \ \mu \leq 0.$$

By the strong duality of LPs, the values of the optimization problems in (7.7) and (7.8) are equivalent implying that if the optimal value of (7.8) is greater than or equal to 0 for a given $x \in \mathbb{R}^n$, then the resulting input u satisfies (7.2). Embedding the conditions imposed by (7.8) as constraints in an optimization problem yields the QP in (7.6). Under the assumption that $K_{\text{rcbf}}(x)$ is nonempty for each $x \in \mathbb{R}^n$, the optimal value of (7.7), and thus of (7.8) by strong duality, is greater than or equal to 0, which implies that (7.6) is feasible for each $x \in \mathbb{R}^n$ and that $k(x) \in K_{\text{rcbf}}(x)$ for each $x \in \mathbb{R}^n$. It then follows from the assumption that the resulting control policy $u = k(x)$ is locally Lipschitz and Lemma 7.1 that such a policy renders $\mathcal{C}$ forward invariant for the closed-loop system, as desired. $\qquad\square$

Remark 7.1 An alternative way to replacing (7.4) with (7.5) would be to use the fact that, for an LP, the optimum value is achieved at a vertex of the feasible set. Therefore, it is possible to replace the constraint given by (7.4) with an enumeration of constraints obtained by replacing θ with each corner of the feasible polyhedron $A\theta \leq b$. In general, however, this would result in a number of constraints that grows combinatorially in the number of half spaces in $A\theta \leq b$. Intuitively, this is avoided in (7.5) because the dual variable μ automatically selects the worst-case corner.

Remark 7.2 Although we have stated all results here for relative degree one CBFs, the same recipe outlined in previous chapters can be used to extend this approach to high order CBFs. Similar to previous chapters, such an extension is contingent on the uncertain parameters θ only appearing in the highest order Lie derivative of h along the dynamics (7.1).

7.1.2   Robust Control Lyapunov Functions

The duality-based approach developed for robust safety naturally extends to robust stabilization problems using the notion of a robust CLF for systems of the form (7.1). For all results in this section we make the following assumption:

Assumption 7.2 The uncertain system (7.1) satisfies $f(0) = 0$ and $\varphi(0, 0) = 0$ so that the origin is an equilibrium point of the unforced system.

Definition 7.2 (*Robust CLF*) A Lyapunov function candidate $V : \mathbb{R}^n \to \mathbb{R}_{\geq 0}$ is said to be a Robust CLF (RCLF) for (7.1) if there exists $\gamma \in \mathcal{K}$ such that for all $x \in \mathbb{R}^n \setminus \{0\}$

$$\inf_{u \in \mathcal{U}} \sup_{\theta \in \Theta} \dot{V}(x, u, \theta) < -\gamma(V(x)), \tag{7.9}$$

where $\dot{V}(x, u, \theta) = L_f V(x) + L_g V(x)u + L_\varphi V(x, u)\theta$.

Now consider the set

$$K_{\mathrm{rclf}}(x) := \{u \in \mathcal{U} \mid L_f V(x) + L_g V(x)u + \sup_{\theta \in \Theta} L_\varphi V(x, u)\theta \leq -\gamma(V(x))\},$$

of all control values satisfying the condition from (7.9). The following lemma shows that any locally Lipschitz controller satisfying the conditions of Definition 7.2 renders the origin asymptotically stable for (7.1).

Lemma 7.2 *If V is a RCLF for* (7.1) *and Assumptions 7.1–7.2 hold, then any control policy* $u = k(x)$, *locally Lipschitz on* $\mathbb{R}^n \setminus \{0\}$, *satisfying* $k(x) \in K_{\mathrm{rclf}}(x)$ *for each* $x \in \mathbb{R}^n$ *renders the origin asymptotically stable for* (7.1).

Proof The derivative of V along the closed-loop system is upper bounded as

$$\begin{aligned}
\dot{V}(x) &= L_f V(x) + L_g V(x)k(x) + L_\varphi V(x, k(x))\theta \\
&\leq L_f V(x) + L_g V(x)k(x) + \sup_{\theta \in \Theta} L_\varphi V(x, k(x))\theta \\
&\leq -\gamma(V(x)),
\end{aligned}$$

and asymptotic stability follows from Theorem 2.2.   □

Following the same duality-based approach as in the previous section we can make the synthesis of robust stabilizing controllers more tractable than as presented in Definition 7.2. The dual of the LP $\sup_{\theta \in \Theta} L_\varphi V(x, u)\theta$ is given by

$$\inf_{\lambda \geq 0} \quad b^\top \lambda \tag{7.10}$$
$$\text{subject to} \quad \lambda^\top A = L_\varphi V(x, u),$$

where λ is the dual variable. This allows to generate inputs satisfying condition (7.9) by solving the following QP:

$$\min_{u \in \mathcal{U}, \lambda \geq 0} \quad \frac{1}{2} \|u\|^2$$

$$\text{subject to} \quad L_f V(x) + L_g V(x)u + b^\top \lambda \leq -\gamma(V(x)) \tag{7.11}$$

$$\lambda^\top A = L_\varphi V(x, u),$$

as shown in the following theorem.

Theorem 7.2 *Let the assumptions of Lemma 7.2 hold. Then, any solution to (7.11), $u = k(x)$, locally Lipschitz on $\mathbb{R}^n \setminus \{0\}$, renders the origin asymptotically stable for the closed-loop system.*

Proof Follows the same steps as that of Theorem 7.1. □

7.2 Online Learning for Uncertainty Reduction

The previous section demonstrates how to robustly account for system uncertainty to guarantee stability and/or safety; however, the initial bounds on the system uncertainty may be highly conservative, which could restrict the system from exploring much of the safe set and, as illustrated in Sect. 7.3, could produce controllers that require large amounts of control effort to enforce stability and safety. A more attractive approach is to leverage input-output data generated by the system at run-time in an effort to identify the system uncertainty, which can be used to reduce the conservatism of the approach outlined in the previous section. Such an approach was leveraged using techniques from adaptive control in previous chapters. Here, we present an alternative approach based on the idea of *set membership identification* (SMID), which is a model identification approach commonly used in the model predictive control literature. Rather than maintaining a point-wise estimate of the uncertain parameters (as in the adaptive control approach), the SMID approach maintains an entire feasible set of parameters. We will effectively use this approach to shrink the hyperrectangle Θ containing θ down to a smaller set to reduce the conservatism of the robust approach presented in the previous section.

Following the approach from Sect. 4.2, let $\Delta t \in \mathbb{R}_{>0}$ be the length of an integration window and note that over any finite time interval $[t - \Delta t, t] \in \mathbb{R}_{\geq 0}$, the Fundamental Theorem of Calculus can be used to represent (7.1) as the linear regression model from (4.18):

$$\mathcal{Y}(t) = \mathcal{F}(t)\theta,$$

where

$$\mathcal{Y}(t) := x(t) - x(t - \Delta t) - \int_{t\max\{t-\Delta t, 0\}}^{t} (f(x(s)) + g(x(s))u(s))\, ds$$

$$\mathcal{F}(t) := \int_{\max\{t-\Delta t, 0\}}^{t} \varphi(x(s), u(s))\, ds. \tag{7.12}$$

Our goal is now to use the above relation to shrink the set of possible parameters Θ using input-output data collected online. To this end, let $\mathcal{H}(t) := \{(\mathcal{Y}_j(t), \mathcal{H}_j(t))\}_{j=1}^{M}$ be a history stack with $M \in \mathbb{N}$ entries. Letting $\{t_k\}_{k \in \mathbb{Z}_{\geq 0}}$ be a strictly increasing sequence of times with $t_0 = 0$, consider the corresponding sequence of sets

$$\Xi_0 = \Theta$$

$$\Xi_k = \{\theta \in \Xi_{k-1} \mid -\varepsilon \mathbf{1}_n \leq \mathcal{Y}_j(t_k) - \mathcal{F}_j(t_k)\theta \leq \varepsilon \mathbf{1}_n, \ \forall j \in \mathcal{M}\},$$

where $\mathbf{1}_n$ is an n-dimensional vector of ones, which is the set of all parameters that approximately satisfy (4.18) for each $j \in \mathcal{M}$ with precision[2] $\varepsilon \in \mathbb{R}_{>0}$. In practice, the set Ξ_k can be computed by solving, for each $i \in \{1, \ldots, p\}$, the pair of LPs

$$\underline{\theta}_i^k = \arg\min_{\theta} \quad \theta_i$$

$$\text{s.t.} \quad \mathcal{Y}_j(t_k) - \mathcal{F}_j(t_k)\theta \leq \varepsilon \mathbf{1}_n \ \forall j$$

$$\mathcal{Y}_j(t_k) - \mathcal{F}_j(t_k)\theta \geq -\varepsilon \mathbf{1}_n \ \forall j \tag{7.13}$$

$$A_{k-1}\theta \leq b_{k-1},$$

$$\overline{\theta}_i^k = \arg\max_{\theta} \quad \theta_i$$

$$\text{s.t.} \quad \mathcal{Y}_j(t_k) - \mathcal{F}_j(t_k)\theta \leq \varepsilon \mathbf{1}_n \ \forall j$$

$$\mathcal{Y}_j(t_k) - \mathcal{F}_j(t_k)\theta \geq -\varepsilon \mathbf{1}_n \ \forall j \tag{7.14}$$

$$A_{k-1}\theta \leq b_{k-1},$$

where θ_i is the ith component of θ and A_{k-1}, b_{k-1} capture the halfspace constraints imposed by Ξ_{k-1}. The updated set of possible parameters is then taken as

$$\Xi_k = [\underline{\theta}_1^k, \overline{\theta}_1^k] \times \cdots \times [\underline{\theta}_p^k, \overline{\theta}_p^k]. \tag{7.15}$$

The following result shows that the true parameters always belong to the set of possible parameters generated by the SMID scheme.

Lemma 7.3 *Provided that Assumption 7.1 holds and the sequence of sets $\{\Xi_k\}_{k \in \mathbb{Z}_{\geq 0}}$ is generated according to (7.13)–(7.15), then $\Xi_k \subseteq \Xi_{k-1} \subseteq \Theta$ and $\theta \in \Xi_k$ for all $k \in \mathbb{Z}_{\geq 0}$.*

---

[2] The constant ε can be seen as a parameter governing the conservativeness of the identification scheme, which can be used to account for disturbances, noise, unmodeled dynamics, and/or numerical integration errors.

Proof The observation that $\Xi_k \subseteq \Xi_{k-1}$ for all $k \in \mathbb{Z}_{\geq 0}$ follows directly from (7.13) and (7.14) since the constraint $A_{k-1}\theta \leq b_{k-1}$ ensures that $\underline{\theta}_i^k, \overline{\theta}_i^k \in [\underline{\theta}_i^{k-1}, \overline{\theta}_i^{k-1}]$ for all i implying $[\underline{\theta}_i^k, \overline{\theta}_i^k] \subseteq [\underline{\theta}_i^{k-1}, \overline{\theta}_i^{k-1}]$ for all i. It then follows from (7.15) and $\Xi_0 = \Theta$ that $\Xi_k \subseteq \Xi_{k-1} \subseteq \Theta$ for all $k \in \mathbb{Z}_{\geq 0}$. Our goal is now to show that $\theta \in \Xi_{k-1} \implies \theta \in \Xi_k$. For any $k \in \mathbb{Z}_{\geq 0}$, relation (4.18) implies that θ belongs to the set

$$H_k = \{\theta \in \mathbb{R}^p \mid \mathcal{Y}_j(t_k) - \mathcal{F}_j(t_k)\theta = 0\}$$

for all $j \in \mathcal{M}$. Additionally, for any $k \in \mathbb{Z}_{\geq 0}$ the constraints in (7.13) and (7.14) ensure that $\Xi_k \subset H_k^- \cap H_k^+$, where

$$H_k^- = \{\theta \in \mathbb{R}^p \mid \mathcal{Y}_j(t_k) - \mathcal{F}_j(t_k)\theta \geq -\varepsilon \mathbf{1}_n\}$$
$$H_k^+ = \{\theta \in \mathbb{R}^p \mid \mathcal{Y}_j(t_k) - \mathcal{F}_j(t_k)\theta \leq \varepsilon \mathbf{1}_n\},$$

for all $j \in \mathcal{M}$. It then follows from $\theta \in H_k$ and $H_k \subset H_k^- \cap H_k^+$ that $\theta \in H_k^- \cap H_k^+$. The last constraint in (7.13) and (7.14) ensures that $\Xi_k \subset H_k^- \cap H_k^+ \cap \Xi_{k-1}$, which implies that $\theta \in \Xi_k$ as long as $\theta \in \Xi_{k-1}$. Since $\theta \in \Xi_0$ it inductively follows from $\theta \in \Xi_{k-1} \implies \theta \in \Xi_k$ for all $k \in \mathbb{Z}_{\geq 0}$ that $\theta \in \Xi_k$ for all $k \in \mathbb{Z}_{\geq 0}$. $\qquad\square$

The following propositions demonstrate that if h and V are a RCBF and RCLF, respectively, for (7.1) with respect to the original parameter set Θ, then they remain so for the parameter sets generated by the SMID algorithm.

Proposition 7.1 *Let h be a RCBF for (7.1) on a set $\mathcal{C} \subset \mathbb{R}^n$ in the sense that there exists $\alpha \in \mathcal{K}_\infty^e$ such that (7.2) holds for all $x \in \mathbb{R}^n$. Provided the assumptions of Lemma 7.3 hold, then*

$$\sup_{u \in \mathcal{U}} \inf_{\theta \in \Xi_k} \dot{h}(x, u, \theta) \geq -\alpha(h(x)),$$

for all $x \in \mathbb{R}^n$ and all $k \in \mathbb{Z}_{\geq 0}$.

Proof Let $\theta_k^* \in \Xi_k$ be the solution to the LP $\inf_{\theta \in \Xi_k} L_\varphi h(x, u)\theta$ for some (x, u). Since $\Xi_{k+1} \subseteq \Xi_k$ by Lemma 7.3 one of the following holds: either (i) $\theta_k^* \in \Xi_{k+1}$ or (ii) $\theta_k^* \in \Xi_k \backslash \Xi_{k+1}$. For case (i), if the infimum is achieved over the set Ξ_{k+1}, then θ_k^* would also be an optimal solution to the LP $\inf_{\theta \in \Xi_{k+1}} L_\varphi h(x, u)\theta$ and

$$\inf_{\theta \in \Xi_{k+1}} L_\varphi h(x, u)\theta = \inf_{\theta \in \Xi_k} L_\varphi h(x, u)\theta.$$

For case (ii) if $\theta_k^* \in \Xi_k \backslash \Xi_{k+1}$, then necessarily

$$\inf_{\theta \in \Xi_{k+1}} L_\varphi h(x, u)\theta \geq \inf_{\theta \in \Xi_k} L_\varphi h(x, u)\theta,$$

otherwise the infimum would have been achieved over Ξ_{k+1} since $\Xi_k \supseteq \Xi_{k+1}$. Thus, since the RCBF condition (7.2) holds over Θ and $\Xi_k \subseteq \Theta$ for all $k \in \mathbb{Z}_{\geq 0}$ by Lemma 7.3, we have

$$\inf_{\theta \in \Xi_k} L_\varphi h(x, u)\theta \geq \inf_{\theta \in \Theta} L_\varphi h(x, u)\theta,$$

for all $k \in \mathbb{Z}_{\geq 0}$. The preceding argument implies

$$\sup_{u \in \mathcal{U}} \inf_{\theta \in \Xi_k} \dot{h}(x, u, \theta) \geq \sup_{u \in \mathcal{U}} \inf_{\theta \in \Theta} \dot{h}(x, u, \theta) \geq -\alpha(h(x)),$$

for all $x \in \mathbb{R}^n$ and $k \in \mathbb{Z}_{\geq 0}$, as desired. $\square$

Proposition 7.2 *Let V be a RCLF for (7.1) in the sense that there exists $\gamma \in \mathcal{K}$ such that (7.9) holds for all $x \in \mathbb{R}^n$. Provided the assumptions of Lemma 7.3 hold, then*

$$\inf_{u \in \mathcal{U}} \sup_{\theta \in \Xi_k} \dot{V}(x, u, \theta) \leq -\gamma(V(x)),$$

for all $x \in \mathbb{R}^n$ and all $k \in \mathbb{Z}_{\geq 0}$.

Proof The proof parallels that of Proposition 7.1. $\square$

7.3   Numerical Examples

Example 7.2 We first consider a scenario for a two-dimensional nonlinear system with naturally unsafe dynamics in the sense that trajectories of the system leave the safe set without intervention from a controller. The dynamics of the system are in the form of (7.1) and are given by $f(x) = g(x) = 0$ such that

$$\begin{bmatrix} \dot{x}_1 \\ \dot{x}_2 \end{bmatrix} = \underbrace{\begin{bmatrix} x_1 & x_2 & 0 & 0 \\ 0 & 0 & x_1^3 & x_2 u \end{bmatrix}}_{\varphi(x,u)} \underbrace{\begin{bmatrix} \theta_1 \\ \theta_2 \\ \theta_3 \\ \theta_4 \end{bmatrix}}_{\theta}.$$

The uncertain parameters are assumed to lie in the set

$$\Theta = [-1.2, 0.2] \times [-2, -0.1] \times [0.5, 1.4] \times [0.8, 1.2] \subset \mathbb{R}^4.$$

The objective is to regulate the system to the origin while remaining in a set $\mathcal{C} \subset \mathbb{R}^2$ characterized as in (3.3) with

$$h(x) = 1 - x_1 - x_2^2.$$

The regulation objective is achieved by considering the RCLF candidate

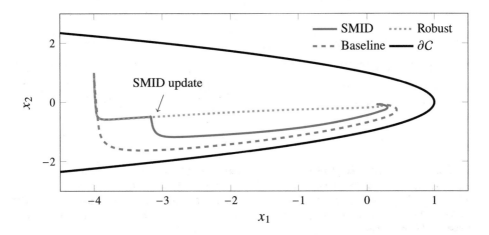

Fig. 7.1 Trajectory of the nonlinear system under various controllers. The solid blue curve depicts the trajectory with SMID, the dotted orange curve depicts the trajectory without SMID, the purple curve illustrates the trajectory under a standard CBF-QP with exact model knowledge, and the black curve denotes the boundary of the safe set

$$V(x) = \tfrac{1}{4}x_1^4 + \tfrac{1}{2}x_2^2,$$

with $\gamma(s) = \tfrac{1}{2}s$ and the safety objective is achieved by considering the RCBF candidate with h as above and $\alpha(s) = s^3$. Given a RCLF, RCBF, and uncertainty set Θ, one can form a QP as noted after Theorem 7.2 to generate a closed-loop control policy that guarantees stability and safety provided the sufficient conditions of Theorems 7.1 and 7.2 are satisfied. To illustrate the impact of the integral SMID procedure, simulations are run with and without SMID active, the results of which are provided in Figs. 7.1 and 7.2. The parameters associated with the SMID simulation are $\Delta t = 0.3$, $\varepsilon = 0.1$, $M = 20$. The M data points in LPs (7.13) and (7.14) are collected using a moving window approach, where the M most recent data points are used to update the uncertainty set. As illustrated in Fig. 7.1 the trajectory under the RCLF-RCBF-QP achieves the stabilization and safety objective with and without SMID; however, the trajectory without any parameter identification is significantly more conservative and is unable to approach the boundary of the safe set. In contrast, the trajectory with SMID is able to approach the boundary of the safe set as more data about the system becomes available. In particular, both trajectories follow an identical path up until $t = \Delta t$, at which point the set of possible parameters is updated, causing the blue curve (SMID) to deviate from the orange curve (no SMID) in Fig. 7.1. In fact, even after the first SMID update the blue curve closely resembles the purple curve, which corresponds to the trajectory under a CBF-QP with perfect model knowledge. Although the parameters have not been exactly identified by the end of the simulation (see Fig. 7.2), the modest reduction in uncertainty offered by the SMID approach greatly reduces the conservatism of the purely robust approach.

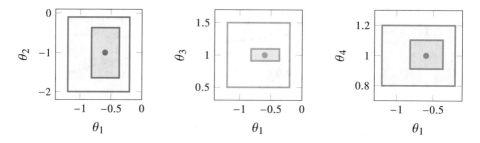

Fig. 7.2 Set-based estimate of the uncertain parameters for the nonlinear system. From left to right, the plots illustrate the uncertainty set Θ projected onto the $\theta_1 \times \theta_2$, $\theta_1 \times \theta_3$, and $\theta_1 \times \theta_4$ axes, respectively. In each plot the pale rectangle represents the original uncertainty set, the dark rectangle represents the final uncertainty set generated by the SMID algorithm, and the dot represents the true values of the parameters

Example 7.3 We now consider a robotic navigation task and demonstrate how to incorporate HOCBFs into the developed framework. The robot is modeled as a planar double integrator with uncertain mass and friction effects of the form

$$m\ddot{q} = u - c\dot{q},$$

where $q \in \mathbb{R}^2$ denotes the robot's position, $m \in \mathbb{R}_{>0}$ its mass, $c = \text{diag}([c_1 \; c_2]) \in \mathbb{R}^{2\times2}$ a vector of friction coefficients, and $u \in \mathbb{R}^2$ its commanded acceleration. Taking the state as $x = [q^\top \; \dot{q}^\top]^\top \in \mathbb{R}^4$ and the uncertain parameters as $\theta = [\frac{c_1}{m} \; \frac{c_2}{m} \; m]^\top \in \mathbb{R}^3$ allows the system to be put into the form of (7.1) as

$$\underbrace{\begin{bmatrix} \dot{x}_1 \\ \dot{x}_2 \\ \dot{x}_3 \\ \dot{x}_4 \end{bmatrix}}_{\dot{x}} = \underbrace{\begin{bmatrix} x_3 \\ x_4 \\ 0 \\ 0 \end{bmatrix}}_{f(x)} + \underbrace{\begin{bmatrix} 0 & 0 & 0 \\ 0 & 0 & 0 \\ -x_3 & 0 & u_1 \\ 0 & -x_4 & u_2 \end{bmatrix}}_{\varphi(x,u)} \underbrace{\begin{bmatrix} \frac{c_1}{m} \\ \frac{c_2}{m} \\ m \end{bmatrix}}_{\theta}.$$

The objective is to drive the robot to the origin while avoiding a circular obstacle of radius $r \in \mathbb{R}_{>0}$ centered at $[x_o \; y_o]^\top \in \mathbb{R}^2$. The state constraint set can be described as the zero superlevel set of

$$h(x) = (x_1 - x_o)^2 + (x_2 - y_o)^2 - r^2,$$

which has relative degree 2 with respect to both the control input and uncertain parameters as required by Remark 7.2.

To further demonstrate the advantage of reducing the level of uncertainty online, we simulate the double integrator under a robust HOCBF-based policy with and without the SMID algorithm running. For each simulation, the uncertain parameters are assumed to

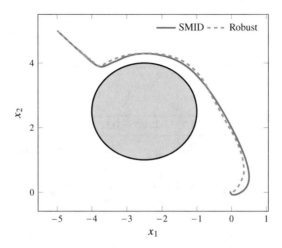

Fig. 7.3 Evolution of the double integrator's position with the SMID algorithm active (blue) and inactive (orange). The gray disk denotes an obstacle of radius $r = 1.5$ centered at $x_o = -2.5, y_o = 2.5$

lie in the set $\Theta = [0, 5] \times [0, 5] \times [0.1, 2]$ and all extended class $\mathcal{K}^e_\infty$ functions used in the HOCBF constraints are chosen as $\alpha(s) = s^3$. The stabilization objective is achieved by considering the same CLF candidate used in Example 5.1 and the controller ultimately applied to the system is computed by filtering the solution to the RCLF-QP (7.11) through a robust HOCBF-QP. The parameters for the SMID algorithm are chosen as $M = 20$, $\Delta t = 0.1$, and $\varepsilon = 1$, where data is recorded using the same technique as in the previous example. The trajectory of the robot's position with and without the SMID algorithm is illustrated in Fig. 7.3, where each trajectory is shown to satisfy the stability and safety objective. Although the trajectories appear very similar, the controller without SMID generates this trajectory with significantly more control effort (see Fig. 7.4). In fact, within the first second of the simulation such a controller requires control effort that is an order of magnitude higher than that of the controller that reduces the uncertainty online to avoid collision with the obstacle.

7.4 Notes

In this chapter, we presented a duality-based approach to robust and data-driven safety-critical control that allows for the synthesis of CBF and CLF-based controllers using quadratic programming (QP). An alternative to the duality-based approach to designing CBF/CLF-based controllers for systems with additive and multiplicative uncertainty, which first appeared in [1], involves reformulating the optimization problem (7.3) as a *second order cone program* (SOCP), which is a convex optimization problem and hence can be solved efficiently in real-time. Converting (7.3) to a SOCP generally involves assuming that the Lie

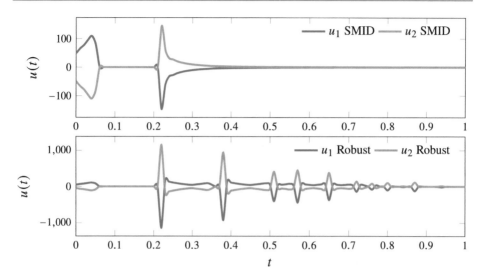

Fig. 7.4 Evolution of the control input for the double integrator with the SMID algorithm active (top) and inactive (bottom)

derivative of h along the dynamics (7.1) can be lower-bounded as

$$\dot{h} = L_f h(x) + L_g h(x)u + L_\varphi h(x, u)\theta$$
$$= L_f h(x) + L_g h(x)u - (a(x) + b(x)\|u\|)\bar{\theta},$$

for some locally Lipschitz functions $a : \mathbb{R}^n \to \mathbb{R}$, $b : \mathbb{R}^n \to \mathbb{R}$ and some known bound $\|\theta\| \le \bar{\theta}$. The above condition is still a nonlinear function of the control input because of the appearance of $\|u\|$; however, it can be recast as a second order cone constraint with an explicit conversion detailed in [2]. Promising works that take the SOCP approach to accounting for model uncertainty include [3–5]. The SOCP approach has also found applications in developing CBF-based controllers that account for measurement uncertainty [2, 6] and input delays [7].

As mentioned in Remark 7.1, it is also possible to convert the optimization problem from (7.3) into a QP by enumerating all the vertices of Θ as constraints - an approach taken in works such as [8, 9]. Although this approach leads to control synthesis using a QP, the number of constraints can grow rapidly in higher dimensions. For example, the vertex representation of a p-dimensional hyperrectangle results in 2^p constraints, whereas the halfspace representation only results in $2p$ constraints. Although the duality-based approach presented in this chapter does not directly correspond to using the halfspace representation of Θ in a QP, it does produce a QP whose number of constraints scale *linearly*, rather than exponentially. The system dynamics and safety constraint used in Example 7.2 are taken from [10].

References

1. Cohen MH, Belta C, Tron R (2022) Robust control barrier functions for nonlinear control systems with uncertainty: a duality-based approach. In: Proceedings of the ieee conference on decision and control, pp 174–179
2. Dean S, Taylor AJ, Cosner R, Recht B, Ames AD (2021) Guaranteeing safety of learned perception modules via measurement-robust control barrier functions. In: Proceedings of the 2020 conference on robot learning, vol 155. Proceedings of machine learning research, pp 654–670
3. Taylor AJ, Dorobantu VD, Dean S, Recht B, Yue Y, Ames AD (2021) Towards robust data driven-control synthesis for nonlinear systems with actuation uncertainty. In: Proceedings of the ieee conference on decision and control, pp 6469–6476
4. Castaneda F, Choi JJ, Zhang B, Tomlin CJ, Sreenath K (2021) Pointwise feasibility of gaussian process-based safety-critical control under model uncertainty. In: Proceedings of the ieee conference on decision and control, pp 6762–6769
5. Dhiman V, Khojasteh MJ, Franceschetti M, Atanasov N (2021) Control barriers in bayesian learning of system dynamics. In: IEEE transactions on automatic control
6. Cosner RK, Singletary AW, Taylor AJ, Molnar TG, Bouman KL, Ames AD (2021) Measurement-robust control barrier functions: Certainty in safety with uncertainty in state. In: Proceedings of the IEEE/RSJ international conference on intelligent robots and systems, pp 6286–6291
7. Molnar TG, Kiss AK, Ames AD, Orosz G (2022) Safety-critical control with input delay in dynamic environment. In: IEEE transactions on control systems technology
8. Dawson C, Qin Z, Gao S, Fan C (2021) Safe nonlinear control using robust neural lyapunov-barrier functions. In: Proceedings of the 5th annual conference on robot learning
9. Emam Y, Glotfelter P, Wilson S, Notomista G, Egerstedt M (2021) Data-driven robust barrier functions for safe, long-term operation. In: IEEE transactions on robotics
10. Jankovic M (2018) Robust control barrier functions for constrained stabilization of nonlinear systems. Automatica 96:359–367

Safe Exploration in Model-Based Reinforcement Learning

In this chapter, we show how to use reinforcement learning (RL) to generate a control policy for the same uncertain dynamical system considered in the previous chapters. Specifically, we present an *online* model-based RL (MBRL) algorithm that balances the often competing objectives of learning and safety by simultaneously learning the value function of an optimal control problem and the uncertain parameters of a dynamical system using real-time data. Central to our approach is a safe exploration framework based on the adaptive control barrier functions introduced in Chap. 5. We start by formulating an infinite-horizon optimal control problem as an RL problem in Sect. 8.1. Approximations for the value function used in the RL algorithm are discussed in Sect. 8.2. The main part of this chapter is Sect. 8.3, where we present the online MBRL algorithm. We illustrate the method with numerical examples in Sect. 8.4 and conclude with final remarks, references, and suggestions for further reading in Sect. 8.5.

Reinforcement learning (RL) is a machine learning technique used to solve sequential decision-making problems in the face of uncertainty via function approximation. The sequential decision-making problem we consider in this chapter is that of optimal control in which the goal is to construct a control policy that optimizes an objective function over a long time horizon, whereas the uncertainty we consider is that of uncertainty in the vector fields governing the system dynamics. Thus, in the context of this book, we simply view RL as a technique used to solve optimal control problems when the system dynamics are uncertain. The typical high-level approach taken in using RL to solve optimal control problems for uncertain system dynamics is as follows:

1. A user specifies a cost function and an initial control policy;
2. The control policy is applied to the dynamical system (either through simulation or experiment) to generate a closed-loop system trajectory;

© The Author(s), under exclusive license to Springer Nature Switzerland AG 2023
M. Cohen and C. Belta, *Adaptive and Learning-Based Control of Safety-Critical Systems*,
Synthesis Lectures on Computer Science,
https://doi.org/10.1007/978-3-031-29310-8_8

3. The data (e.g., state trajectory, control trajectory, total cost, etc.) from such a trajectory is used to update the control policy offline in an attempt to improve the policy with respect to the cost function;
4. Steps 2 and 3 are repeated until the user is satisfied with the control policy.

The above description is, of course, an extremely simplified depiction of the RL pipeline, but captures the main essence of many standard RL techniques. Where many RL techniques differ is in how the trajectory data is used to improve the policy. For example *model free* RL does not assume any knowledge of the vector fields governing the system dynamics and, instead, updates the policy directly from the observed trajectory data. An argument in favor of model-free RL is that it may be more challenging to construct a complicated dynamics model than simply learning a policy that maps states to control actions. By not explicitly[1] using model knowledge in the control design, such approaches may also generalize well beyond their training data. The drawback of model-free RL is that it is generally extremely sample inefficient, and therefore may require generating many trajectories until convergence to a suitable policy is obtained. On the other hand, *model-based* RL (MBRL) approaches often use trajectory data to learn a dynamics model, which is then exploited in the process of learning a better policy. Such approaches are often considered to be more sample efficient, requiring less data before a suitable policy is obtained. An argument against MBRL is that the resulting policy may be heavily biased towards the learned model, which, if not representative of the true underlying dynamics, may perform poorly when deployed on a system.

Regardless of the method that is ultimately used to update the policy, we refer to the above paradigm as *episodic* or *offline* RL as one must run multiple trials or episodes to obtain data, and the actual learning (i.e., updating the policy) is generally not done in real-time as the system is being controlled. This offline approach has demonstrated a tremendous amount of success in low-risk settings where generating an abundance of data can be done without real-world consequences, but presents a number of challenges in the context of safety-critical systems. These challenges mainly stem from the idea that learning and safety seem to be fundamentally at odds with each other: learning in a RL context requires exploring a variety of different actions to generate sufficiently rich data whereas enforcing safety requires restricting actions to those that can only be certified as safe. The fact that a variety of (potentially unsafe) actions may need to be explored before convergence to a suitable policy is obtained using the aforementioned episodic approach may not be a limiting factor in simulation; however, such an approach significantly limits the applicability of these ideas to safety-critical systems where failures during training are deemed to be unacceptable. These challenges can be partially mitigated if one has unlimited access to an accurate simulator of the system under consideration, where mistakes made during learning are largely inconsequential; however, if such a simulator is not fully representative of the physical

---

[1] Of course, any policy learned from simulation data will implicitly depend on the underlying model used to generate such data.

system, then policies trained on data generated by such a simulator may yield undesirable performance when deployed on the physical system.

The aforementioned challenges motivate the development of RL algorithms that operate *online* in which real-time data generated by the system is used to learn a desirable policy and adapt to situations that may be difficult to account for using a policy trained using purely offline data. In the present chapter we introduce an *online* MBRL algorithm that balances the often competing objectives of learning and safety by simultaneously learning the value function of an optimal control problem and the uncertain parameters of a dynamical system safely using real-time data. In particular, we develop a safe exploration framework in which the learned system dynamics are used to simulate on-the-fly exploratory actions needed to generate sufficiently rich data for learning while simultaneously shielding the learning policy from unsafe actions on the physical system using the adaptive control barrier functions introduced in Chap. 5.

8.1   From Optimal Control to Reinforcement Learning

In this chapter, we once again consider the nonlinear control affine system with parametric uncertainty (4.1) with dynamics

$$\dot{x} = f(x) + F(x)\theta + g(x)u.$$

As in previous chapters, our main objective is to design an adaptive controller that stabilizes the origin of (4.1) while satisfying some safety criteria in the sense that closed-loop trajectories should remain in the set (3.3):

$$C = \{x \in \mathbb{R}^n \mid h(x) \geq 0\},$$

at all times. Rather than accomplishing the stabilization objective by constructing a control Lyapunov function (CLF), we seek a control policy that minimizes the user-specified infinite-horizon cost functional

$$J(x_0, u(\cdot)) = \int_0^\infty \ell(x(s), u(s))ds, \tag{8.1}$$

where $\ell : \mathbb{R}^n \times \mathbb{R}^m \to \mathbb{R}_{\geq 0}$ is the running cost, assumed to take the form

$$\ell(x, u) = Q(x) + u^\top Ru, \tag{8.2}$$

where $Q : \mathbb{R}^n \to \mathbb{R}_{\geq 0}$ is continuously differentiable and positive definite, and $R \in \mathbb{R}^{m \times m}$ is symmetric and positive definite. Framing the control objective as the optimization of a cost functional provides a natural way of encoding desired performance specifications that may be challenging to encode when constructing a CLF.

Solutions to the optimal control problem described by the infinite-horizon cost (8.1) are typically characterized in terms of the *optimal value function* or *cost-to-go*

$$V^*(x) = \inf_{u(\cdot)} J(x, u(\cdot)). \tag{8.3}$$

Note that the above minimization is an infinite-dimensional optimization performed over the space of control functions $u(\cdot)$. For the remainder of our development, we impose the following restriction on the value function.

Assumption 8.1 The optimal value function (8.3) is continuously differentiable, and its gradient is locally Lipschitz.

Provided the above assumption holds, the value function can be shown to be the unique positive definite solution to the *Hamilton–Jacobi–Bellman* (HJB) partial differential equation

$$0 = \inf_{u \in \mathbb{R}^m} H(x, u, \nabla V^*(x)), \quad \forall x \in \mathbb{R}^n, \tag{8.4}$$

with a boundary condition of $V^*(0) = 0$, where

$$H(x, u, \lambda) := \lambda^\top (f(x) + F(x)\theta + g(x)u) + \ell(x, u) \tag{8.5}$$

is the Hamiltonian, and $\lambda \in \mathbb{R}^n$ is the costate vector. Provided there exists a continuously differentiable positive definite function V^* satisfying the HJB, taking the minimum on the right-hand-side of (8.4) yields the optimal feedback control policy as

$$k^*(x) = -\frac{1}{2}R^{-1}L_g V^*(x)^\top. \tag{8.6}$$

The following fundamental result illustrates that any positive definite and continuously differentiable function satisfying the HJB is also a CLF.

Theorem 8.1 (Value functions are CLFs) *Under Assumption 8.1, the value function V^* is a Lyapunov function for the closed-loop system (4.1) equipped with the optimal feedback controller $u = k^*(x)$, which asymptotically stabilizes the closed-loop system to the origin.*

Proof We first note that the value function is positive definite by construction based on (8.1) and (8.2). Substituting the optimal policy from (8.6) back into the HJB (8.4) yields

$$0 = L_f V^*(x) + L_F V^*(x)\theta + L_g V^*(x)k^*(x) + \ell(x, k^*(x)).$$

Taking the Lie derivative of V^* along the closed-loop vector field and bounding using the above relation yields

$$\dot{V}^*(x) = L_f V^*(x) + L_F V^*(x)\theta + L_g V^*(x)k^*(x) = -\ell(x, k^*(x)) \leq -Q(x), \tag{8.7}$$

which implies V^* is a Lyapunov function for the closed-loop system, and consequently, that the origin is asymptotically stable for the closed-loop system.   □

Although the above result is appealing from a theoretical standpoint, it is of little practical use since solving the HJB for the value function is a much more difficult problem than constructing a CLF. The difficulty in solving the HJB generally arises from the fact that (8.4) does not admit closed-form solutions, except in special cases. For example, when the dynamics are linear and the cost is quadratic in the state and control, the HJB reduces to the algebraic Riccati equation which can be solved easily either in closed-form or numerically. Unfortunately, for nonlinear systems and/or nonquadratic cost functions such closed-form solutions rarely exist and numerical approaches, typically based upon discretization of the time domain, state space, and control space, tend to suffer from the well-known "curse of dimensionality." Moreover, even if computationally efficient solutions to the HJB were available, the fact that the dynamics (4.1) under consideration are partially unknown makes it challenging to guarantee that offline solutions obtained using a possibly inaccurate model will yield desirable behavior when deployed in the actual system.

One way to overcome these challenges is through the use of function approximation in which the value function and control policy are parameterized using a suitable class of function approximators whose parameters are then updated to optimize a performance metric/loss function. In what follows, we demonstrate how the adaptive control tools developed in earlier chapters can be used to develop update laws for the parameters of such function approximators. Taking this approach allows for learning the value function and policy *online* in real-time (i.e., one-shot learning), rather than episodically as it typical in RL approaches, and allows for making guarantees on convergence of the function approximation, and ultimately, stability of the closed-loop system.

8.2   Value Function Approximation

Given that the optimal value function is difficult to compute in general, we seek a parametric approximation of V^* over some compact set $X \subset \mathbb{R}^n$ containing the origin. In order to derive convergence results for our approximations, we limit ourselves to approximation architectures that are linear in the trainable parameters. This restriction prohibits the use of powerful function approximators such as deep neural networks (DNNs) and places the burden of choosing relevant features for function approximation on the user. That is, we assume that over a given compact set X, the value function can be represented as

$$V^*(x) = W^\top \phi(x) + \varepsilon(x), \tag{8.8}$$

where $W \in \mathbb{R}^l$ is a vector of unknown ideal weights, $\phi : \mathbb{R}^n \to \mathbb{R}^l$ is a continuously differentiable feature vector, and $\varepsilon : \mathbb{R}^n \to \mathbb{R}$ is the unknown continuously differentiable function reconstruction error. The assumption that V^* can be represented as in (8.8) is justified by the universal function approximation theorem (see the notes in Sect. 8.5), which states that, given a continuously differentiable function V^*, a compact set X, and a constant $\bar{\varepsilon}$, there exists a continuously differentiable feature vector ϕ, and a vector of weights W such that V^* and its gradient can be $\bar{\varepsilon}$-approximated over X in the sense that

$$\sup_{x \in X} \left\{ \|V^*(x) - W^\top \phi(x)\| + \|\nabla V^*(x) - W^\top \tfrac{\partial \phi}{\partial x}(x)\| \right\} \leq \bar{\varepsilon}.$$

The universal function approximation theorem, however, does not state how the features should be chosen or how many features are necessary to achieve a desired approximation accuracy over the domain of interest. Despite this, the fact that the value function is a CLF provides guidance towards what features may be relevant to produce an adequate approximation of the value function. Common choices include polynomial features, radial basis functions or other kernel functions, and pretrained DNNs with tunable outer-layer weights.

Given that the weights in (8.8) are unknown, we develop an approximation of the value function by replacing W with an estimate, denoted by $\hat{W}_c \in \mathbb{R}^l$, yielding the approximated value function

$$\hat{V}(x, \hat{W}_c) = \hat{W}_c^\top \phi(x),$$
$$\frac{\partial \hat{V}}{\partial x}(x, \hat{W}_c) = \hat{W}_c^\top \tfrac{\partial \phi}{\partial x}(x). \tag{8.9}$$

For reasons that we will discuss shortly, we maintain a separate approximation of the ideal weights for use in the approximated optimal policy, denoted by $\hat{W}_a \in \mathbb{R}^l$, as

$$\hat{k}(x, \hat{W}_a) = -\frac{1}{2} R^{-1} L_g \hat{V}(x, \hat{W}_a)^\top. \tag{8.10}$$

Given the approximated value function and control policy from (8.9) and (8.10), respectively, our objective is now to develop a performance metric (i.e., "loss function" if one prefers machine learning terminology) that quantifies the accuracy of such approximations. To this end we first define

$$\tilde{W}_c := W - \hat{W}_c$$
$$\tilde{W}_a := W - \hat{W}_a, \tag{8.11}$$

as the weight estimation errors. Since W is unknown we cannot simply compute $\tilde{W}_c$ and $\tilde{W}_a$ and directly use such quantities to construct a meaningful performance metric. Instead, we seek an indirect performance metric related to the quality of the approximations. Such a development will be centered around the HJB (8.4), which provides a necessary and sufficient condition for optimality. In particular, the HJB states that, for any given $x \in \mathbb{R}^n$, the optimal value function and control policy satisfy the relation

$$H(x, k^*(x), V^*(x)) = 0. \tag{8.12}$$

Thus, we take as our performance metric the difference between the Hamiltonian evaluated with the approximated value function and approximated optimal policy and the optimal Hamiltonian as

$$\delta(x, \hat{W}_c, \hat{W}_a) := H(x, \hat{k}(x, \hat{W}_a), \nabla \hat{V}(x, \hat{W}_c)) - H(x, k^*(x), \nabla V^*(x))$$

$$= H(x, \hat{k}(x, \hat{W}_a), \nabla \hat{V}(x, \hat{W}_c)), \qquad (8.13)$$

where the second equality follows from (8.12). We refer to the above quantity as the *Bellman error* (BE), which can be computed at any given state $x \in X$ and for any given estimates $(\hat{W}_c, \hat{W}_a) \in \mathbb{R}^l \times \mathbb{R}^l$ provided the model parameters θ are known, and provides an indirect metric related to the "distance" between the optimal solution and current approximations.

If the model parameters are unknown, however, the Hamiltonian cannot be computed exactly, and we must instead work with an *approximate* Hamiltonian corresponding to an estimate of the uncertain parameters $\hat{\theta} \in \mathbb{R}^p$ as

$$\hat{H}(x, u, \lambda, \hat{\theta}) := \lambda^\top (f(x) + F(x)\hat{\theta} + g(x)u) + \ell(x, u). \qquad (8.14)$$

Following the same steps as before, we can then define an approximated version of the BE as

$$
\begin{aligned}
\hat{\delta}(x, \hat{W}_c, \hat{W}_a, \hat{\theta}) &:= \hat{H}(x, \hat{k}(x, \hat{W}_a), \nabla \hat{V}(x, \hat{W}_c), \hat{\theta}) - H(x, k^*(x), \nabla V^*(x)) \\
&= \hat{H}(x, \hat{k}(x, \hat{W}_a), \nabla \hat{V}(x, \hat{W}_c), \hat{\theta}) \\
&= \nabla \hat{V}(x, \hat{W}_c)^\top (f(x) + F(x)\hat{\theta} + g(x)\hat{k}(x, \hat{W}_a)) + \ell(x, \hat{k}(x, \hat{W}_a)) \\
&= \hat{W}_c^\top \tfrac{\partial \phi}{\partial x}(x)(f(x) + F(x)\hat{\theta} + g(x)\hat{k}(x, \hat{W}_a)) + \ell(x, \hat{k}(x, \hat{W}_a)) \\
&= \hat{W}_c^\top \omega(x, \hat{W}_a, \hat{\theta}) + \ell(x, \hat{k}(x, \hat{W}_a)) \\
&= \omega(x, \hat{W}_a, \hat{\theta})^\top \hat{W}_c + \ell(x, \hat{k}(x, \hat{W}_a)),
\end{aligned}
$$

$$(8.15)$$

where we have defined

$$\omega(x, \hat{W}_a, \hat{\theta}) := \tfrac{\partial \phi}{\partial x}(x)(f(x) + F(x)\hat{\theta} + g(x)\hat{k}(x, \hat{W}_a)), \qquad (8.16)$$

to make the affine dependence of $\hat{\delta}$ on $\hat{W}_c$ explicit. Note that $\hat{\delta}$ is an affine function of $\hat{W}_c$ but a nonlinear function of $\hat{W}_a$. This affine dependence on $\hat{W}_c$ is the motivation for maintaining separate approximations of the ideal weights in the value function and policy as this makes minimizing a performance metric based on $\hat{\delta}^2$ much easier. Similar to (8.13), the approximate BE (8.15) can be computed for any $x \in X$ given an estimate of the model parameters $\hat{\theta}$ and weights $(\hat{W}_c, \hat{W}_a)$ to indirectly quantify the performance of such weight estimates in terms of approximation of the optimal value function and policy. Our objective is then to select the weights $(\hat{W}_c, \hat{W}_a)$ to minimize the total squared BE over the approximation domain

$$\int_X \hat{\delta}^2(x, \hat{W}_c, \hat{W}_a, \hat{\theta}) dx, \qquad (8.17)$$

which we replace with the more tractable minimization of

$$\sum_{i=1}^{N} \hat{\delta}^2(x_i, \hat{W}_c, \hat{W}_a, \hat{\theta}), \tag{8.18}$$

where $\{x_i\}_{i=1}^{N}$ is a collection of $N \in \mathbb{N}$ points sampled from $\mathcal{X}$. For a fixed $\hat{\theta}$ such a minimization can be performed simply by sampling states from $\mathcal{X}$; however, the resulting weight estimates $(\hat{W}_c, \hat{W}_a)$ would be highly biased towards the estimated model parameters $\hat{\theta}$, which may be inaccurate. Hence, to obtain an accurate approximation of the value function, it is necessary to obtain a better estimate of the uncertain model parameters, which can be done using the adaptive control methods outlined in previous chapters. In the following section we provide a pathway towards accomplishing this objective using a model-based reinforcement learning (MBRL) approach that allows for learning the value function, control policy, and uncertain model parameters all simultaneously online.

8.3   Online Model-Based Reinforcement Learning

In this section we introduce a safe exploration framework that allows for jointly learning online the uncertain system dynamics and the optimal value function and policy while guaranteeing safety. Our safe exploration architecture consists of two main components. The first is an adaptive CBF scheme, similar to those outlined in the preceding chapters, which allows for identifying the uncertain parameters θ online while guaranteeing safety. The second is a safe exploration framework that leverages the learned model to simulate and explore potentially unsafe actions to generate data for learning the value function without risking safety violation of the physical system.

8.3.1   System Identification

The first component of our MBRL architecture is a system identification method for learning the uncertain model parameters online, which we accomplish using the concurrent learning technique introduced in previous chapters. To this end, recall that integrating the dynamics (4.1) over some finite time interval yields the linear regression equation from (4.18)

$$\mathcal{Y}(t) = \mathcal{F}(t)\theta,$$

where $\mathcal{Y}$ and $\mathcal{F}$ are defined as in (4.17). Given a history stack of input-output data, we propose to update the parameters according to (5.25) as

$$\dot{\hat{\theta}} = \gamma \sum_{j=1}^{M} \mathcal{F}_j^\top \left(\mathcal{Y}_j - \mathcal{F}_j \hat{\theta} \right),$$

where $\gamma \in \mathbb{R}_{>0}$ is a learning gain and $\mathcal{H} = \{(\mathcal{Y}_j, \mathcal{F}_j)\}_{j=1}^{M}$ is a history stack. For the results in this chapter, we assume that the history stack used in the above update law satisfies the finite excitation condition (see Definition 4.5) so that the parameter estimates exponentially converge to their true values.

Assumption 8.2 The history stack $\mathcal{H}$ used in the update law (5.25) satisfies the finite excitation condition (Definition 4.5), which, by Theorem 4.3, ensures the existence of a Lyapunov-like function $V_\theta : \mathbb{R}^p \times \mathbb{R}_{\geq 0} \to \mathbb{R}_{\geq 0}$ satisfying

$$c_1 \|\tilde{\theta}\|^2 \leq V_\theta(\tilde{\theta}, t) \leq c_2 \|\tilde{\theta}\|^2, \quad \forall (\tilde{\theta}, t) \in \mathbb{R}^p \times \mathbb{R}_{\geq 0}, \tag{8.19a}$$

$$\dot{V}_\theta(\tilde{\theta}, t) \leq -c_3 \|\tilde{\theta}\|^2, \quad \forall (\tilde{\theta}, t) \in \mathbb{R}^p \times \mathbb{R}_{\geq 0}, \tag{8.19b}$$

for some positive constants $c_1, c_2, c_3 \in \mathbb{R}_{>0}$.

Remark 8.1 Although Theorem 4.3 does not directly assert the existence of a Lyapunov-like function satisfying (8.19b) for all $t \in \mathbb{R}_{\geq 0}$, it does establish an exponential bound on the parameter estimation error, which, by standard converse Lyapunov theorems, can be used to show the existence of a Lyapunov-like function satisfying Assumption 8.2.

8.3.2   Safe Exploration via Simulation of Experience

We present in this section our safe exploration method for learning jointly online the value function, control policy, and uncertain system dynamics. Recall that our objective is to design a control policy that optimizes the cost functional from (8.1) while guaranteeing forward invariance of a set $C \subset \mathbb{R}^n$ as defined as the zero superlevel set of a continuously differentiable function $h : \mathbb{R}^n \to \mathbb{R}$. We accomplish the safety objective using a robust adaptive control barrier function (RaCBF) as introduced in Chap. 5.2. In particular, under the assumption that h is a RaCBF (see Definition 5.2), we shield the learned policy $\hat{k}$ from (8.10) using the optimization-based controller

$$k(x, \hat{\theta}, \hat{W}_a) = \underset{u \in \mathbb{R}^m}{\operatorname{argmin}} \; \tfrac{1}{2} \|u - \hat{k}(x, \hat{W}_a)\|^2$$

$$\text{s.t.} \quad L_f h(x) + L_F h(x)\hat{\theta} + L_g h(x)u \geq -\alpha(h(x)) + \|L_F h(x)\|\tilde{\vartheta}, \tag{8.20}$$

where $\tilde{\vartheta}$ is a bound on the parameter estimation error from Assumption 5.2. Although the above policy guarantees safety by Theorem 5.2, it may restrict the system from taking actions that are necessary to generate sufficiently rich data for learning. On the other hand, directly applying the learned policy from (8.10) to the system may not enforce forward invariance of C, which, in our safety-critical setting, would be unacceptable.

Our safe exploration framework is facilitated by the observation that the approximate BE (8.15) need not be evaluated with the policy that is ultimately applied to the system. That is, one can leverage the learned policy (8.10), which may be unsafe, to generate sufficiently rich data for learning the value function while simultaneously shielding such a policy using the RaCBF safety filter (8.20) to prevent the actual system from violating safety-critical constraints. This idea manifests itself as *simulation of experience* in which the learned model and policy are used to simulate potentially unsafe actions that may be beneficial for learning, but that also may be unsafe. To this end, recall from the previous section that our learning objective can be accomplished by minimizing the squared approximate BE over the approximation domain $\mathcal{X}$. To develop a learning algorithm that minimizes such a performance metric, we introduce the normalized loss function

$$L(\hat{W}_c, \hat{W}_a, \hat{\theta}) := \frac{1}{N} \sum_{i=1}^{N} \frac{\hat{\delta}^2(x_i, \hat{W}_c, \hat{W}_a, \hat{\theta})}{2\rho^2(x_i, \hat{W}_a, \hat{\theta})}, \tag{8.21}$$

where

$$\rho(x, \hat{W}_a, \hat{\theta}) := 1 + \omega(x_i, \hat{W}_a, \hat{\theta})^\top \omega(x_i, \hat{W}_a, \hat{\theta}),$$

is a normalization term. The derivative of L with respect to $\hat{W}_c$ is

$$\frac{\partial L}{\partial \hat{W}_c}(\hat{W}_c, \hat{W}_a, \hat{\theta}) = \frac{1}{N} \sum_{i=1}^{N} \frac{(\hat{W}_c^\top \omega(x_i, \hat{W}_a, \hat{\theta}) + \ell(x_i, \hat{k}(x_i, \hat{W}_a)))\omega(x_i, \hat{W}_a, \hat{\theta})^\top}{\rho^2(x_i, \hat{W}_a, \hat{\theta})}$$

$$= \frac{1}{N} \sum_{i=1}^{N} \frac{\omega(x_i, \hat{W}_a, \hat{\theta})^\top}{\rho^2(x_i, \hat{W}_a, \hat{\theta})} \hat{\delta}(x_i, \hat{W}_c, \hat{W}_a, \hat{\theta}). \tag{8.22}$$

Using the above, we update the value function weights using a normalized version of the recursive least squares (RLS) with forgetting/discount factor used in Sect. 6.4 as

$$\dot{\hat{W}}_c = -\Gamma \frac{\kappa_c}{N} \sum_{i=1}^{N} \frac{\omega(x_i, \hat{W}_a, \hat{\theta})}{\rho^2(x_i, \hat{W}_a, \hat{\theta})} \hat{\delta}(x_i, \hat{W}_c, \hat{W}_a, \hat{\theta}), \tag{8.23}$$

$$\dot{\Gamma} = \beta\Gamma - \Gamma \left(\frac{\kappa_c}{N} \sum_{i=1}^{N} \frac{\omega(x_i, \hat{W}_a, \hat{\theta})\omega(x_i, \hat{W}_a, \hat{\theta})^\top}{\rho^2(x_i, \hat{W}_a, \hat{\theta})} \right) \Gamma, \tag{8.24}$$

where $\kappa_c \in \mathbb{R}_{>0}$ is a learning gain and $\beta \in \mathbb{R}_{>0}$ is the discount factor. Based on the proceeding convergence analysis, we update the control policy weights as

$$\dot{\hat{W}}_a = -\kappa_{a_1}(\hat{W}_a - \hat{W}_c) - \kappa_{a_2}\hat{W}_a + \frac{\kappa_c}{N} \sum_{i=1}^{N} \frac{G_\phi(x_i)\hat{W}_a\omega(x_i, \hat{W}_a, \hat{\theta})^\top}{4\rho^2(x_i, \hat{W}_a, \hat{\theta})} \hat{W}_c, \tag{8.25}$$

where

$$G_\phi(x) := \frac{\partial \phi}{\partial x}(x) G_R(x) \frac{\partial \phi}{\partial x}(x)^\top,$$

$$G_R(x) := g(x) R^{-1} g(x)^\top.$$

Remark 8.2   Although the update law (8.25) is helpful in establishing theoretical convergence results, the much simpler update law

$$\dot{\hat{W}}_a = \text{proj}_{\mathcal{W}}(-\kappa_a(\hat{W}_a - \hat{W}_c)), \tag{8.26}$$

where $\text{proj}_{\mathcal{W}}(\cdot)$ is a projection operator that keeps the weights within a convex compact set $\mathcal{W} \subset \mathbb{R}^l$, tends to work well in practice. References for such a projection operator will be provided in the notes for this chapter.

The following lemma places bounds on the least-squares matrix Γ and will play an important role in ensuring convergence of the value function and control policy weights.

Lemma 8.1   *Let* $t \mapsto \hat{W}_c(t)$, $\hat{W}_a(t)$, $\Gamma(t)$, $\hat{\theta}(t)$ *be trajectories generated by the update laws in* (8.23), (8.24), (8.25), *and* (5.25). *Suppose that* $\lambda_{\min}(\Gamma(0)) > 0$ *and that the constant*

$$\lambda_c := \inf_{t \in \mathbb{R}_{\geq 0}} \left\{ \lambda_{\min} \left(\frac{1}{N} \sum_{i=1}^{N} \frac{\omega(x_i, \hat{W}_a(t), \hat{\theta}(t)) \omega(x_i, \hat{W}_a(t), \hat{\theta}(t))^\top}{\rho^2(x_i, \hat{W}_a(t), \hat{\theta}(t))} \right) \right\}, \tag{8.27}$$

is strictly positive. Then, there exist positive constants $\bar{\Gamma}, \underline{\Gamma} \in \mathbb{R}_{>0}$ *such that* $I_L \underline{\Gamma} \leq \Gamma(t) \leq I_L \bar{\Gamma}$ *for all* $t \in \mathbb{R}_{\geq 0}$.

The condition in (8.27) is similar to the concurrent learning conditions needed for parameter convergence in earlier chapters, and serves a similar purpose here as satisfaction of such a condition will allow us to make guarantees about convergence of the weight estimates $(\hat{W}_c, \hat{W}_a)$. Although the minimum eigenvalue in (8.27) can be computed at any given point in time, verifying that the overall condition holds is challenging as this requires reasoning about the future evolution of the weight estimates. We note that such a condition can be heuristically satisfied by densely sampling $\mathcal{X}$ (i.e., choosing a large number N of distinct extrapolation points). Given the preceding lemma we now have all the tools in place to establish convergence guarantees for the weight estimates and stability of the closed-loop system. We first show that provided the hypothesis of Lemma 8.1 and the constant λ_c is sufficiently large, then all estimated parameters converge to a neighborhood of their true values. Before stating the result, for convenience we introduce the notation

$$\overline{\|f(x)\|}_{\mathcal{X}} := \sup_{x \in \mathcal{X}} \|f(x)\|$$

for any continuous $f : \mathcal{X} \to \mathbb{R}^q$ and $q \in \mathbb{N}$.

Theorem 8.2 *Let $z := [\tilde{W}_c^\top \ \tilde{W}_a^\top \ \tilde{\theta}^\top]^\top \in \mathbb{R}^{2l+p}$ be a composite vector of estimation errors and suppose the estimated weights and parameters are updated according to (8.23), (8.24), (8.25), and (5.25). Provided the conditions of Lemma 8.1 are satisfied, Assumption 8.2 holds, and $\lambda_{\min}(M) > 0$, where*

$$
M = \begin{bmatrix}
\frac{\kappa_c \bar{\lambda}}{4} & -\frac{v_{ac}}{2} & -\frac{v_{c\theta}}{2} \\
-\frac{v_{ac}}{2} & \frac{\kappa_{a_1} + \kappa_{a_2}}{4} - v_a & 0 \\
-\frac{v_{c\theta}}{2} & 0 & \frac{c_3}{2}
\end{bmatrix},
\tag{8.28}
$$

and

$$
\bar{\lambda} := \frac{\beta}{2\kappa_c \bar{\Gamma}} + \frac{\lambda_c}{2},
$$

$$
v_{ac} := \kappa_{a_1} + \frac{3\sqrt{3}\kappa_c}{64} \|W\| \overline{\|G_\phi(x)\|}_\chi,
$$

$$
v_{c\theta} := \frac{3\sqrt{3}\kappa_c}{16} \|W\| \overline{\left\|\frac{\partial \phi}{\partial x}(x)F(x)\right\|}_\chi,
\tag{8.29}
$$

$$
v_a := \frac{3\sqrt{3}\kappa_c}{64} \|W\| \overline{\|G_\phi(x)\|}_\chi,
$$

then all estimated parameters exponentially converge to a neighborhood of their true values in the sense that for all $t \in \mathbb{R}_{\geq 0}$

$$
\|z(t)\| \leq \sqrt{\frac{\kappa_2}{\kappa_1} \|z(0)\|^2 e^{-\frac{\kappa_3}{\kappa_2}t} + (1 - e^{-\frac{\kappa_3}{\kappa_2}t})\frac{\iota}{\kappa_1 \kappa_3}},
\tag{8.30}
$$

where

$$
\kappa_1 := \min\left\{\frac{1}{2\bar{\Gamma}}, \frac{1}{2}, c_1\right\},
$$

$$
\kappa_2 := \max\left\{\frac{1}{2\underline{\Gamma}}, \frac{1}{2}, c_2\right\},
$$

$$
\kappa_3 := \min\left\{\frac{\kappa_c \bar{\lambda}}{4}, \frac{\kappa_{a_1} + \kappa_{a_2}}{4}, \frac{c_3}{2}\right\},
\tag{8.31}
$$

$$
\iota := \frac{\iota_c^2}{2\bar{\lambda}\kappa_c} + \frac{\iota_a^2}{2(\kappa_{a_1} + \kappa_{a_2})},
$$

$$
\iota_c := \frac{3\sqrt{3}\kappa_c}{16} \overline{\|\Delta(x)\|}_\chi,
$$

$$
\iota_a := \kappa_{a_2}\|W\| + \frac{3\sqrt{3}\kappa_c}{64} \|W\|^2 \overline{\|G_\phi(x)\|}_\chi.
$$

Before presenting the proof, we aim to provide some intuition to the sufficient conditions of the above theorem and the bound in (8.30). For the condition $\lambda_{\min}(M)$ to hold, the constant λ_c from Lemma 8.1 must be sufficiently large, implying that the data generated through sampling must be sufficiently rich, which can be achieved by more densely sampling χ.

The bound in (8.30) implies that all estimation errors exponentially decay to some ultimate bound at a rate determined by κ_3. The size of this bound is determined by the constant ι, which is dependent on the function reconstruction error ε. Generally, choosing a more expressive basis for V^* will decrease ε and therefore the ultimate bound.

Proof The proof is largely an exercise in bookkeeping: we perform some straightforward, but tedious, algebraic manipulations and then propose a Lyapunov function that certifies the stability of the weight estimates via the comparison lemma. We begin by deriving an alternate form of the approximate BE (8.15). We first note that the approximate Hamiltonian can be expressed as

$$\hat{H} = -\omega(x, \hat{W}_a, \hat{\theta})^\top \tilde{W}_c + W^\top \omega(x, \hat{W}_a, \hat{\theta}) + \ell(x, \hat{k}(x, \hat{W}_a)). \tag{8.32}$$

The second term in (8.32) can be expressed as

$$
\begin{aligned}
W^\top \omega(x, \hat{W}_a, \hat{\theta}) &= W^\top \tfrac{\partial \phi}{\partial x} f + W^\top \tfrac{\partial \phi}{\partial x} F \hat{\theta} + W^\top \tfrac{\partial \phi}{\partial x} g \hat{k} \\
&= W^\top \tfrac{\partial \phi}{\partial x} f + W^\top \tfrac{\partial \phi}{\partial x} F \hat{\theta} - \tfrac{1}{2} W^\top G_\phi \hat{W}_a \\
&= W^\top \tfrac{\partial \phi}{\partial x} f + W^\top \tfrac{\partial \phi}{\partial x} F \hat{\theta} - \tfrac{1}{2} W^\top G_\phi W + \tfrac{1}{2} W^\top G_\phi \tilde{W}_a,
\end{aligned}
$$

where functional arguments are omitted for ease of readability. Similarly, the third term in (8.32) can be expressed as

$$
\begin{aligned}
\ell(x, \hat{k}(x, \hat{W}_a)) &= Q + \hat{k}^\top R \hat{k} \\
&= Q + \left(-\tfrac{1}{2} R^{-1} g^\top \tfrac{\partial \phi}{\partial x}^\top \hat{W}_a\right)^\top R \left(-\tfrac{1}{2} R^{-1} g^\top \tfrac{\partial \phi}{\partial x}^\top \hat{W}_a\right)^\top \\
&= Q + \tfrac{1}{4} \hat{W}_a^\top G_\phi \hat{W}_a \\
&= Q + \tfrac{1}{4} W^\top G_\phi \hat{W}_a - \tfrac{1}{4} \tilde{W}_a^\top G_\phi \hat{W}_a \\
&= Q + \tfrac{1}{4} W^\top G_\phi W - \tfrac{1}{4} W^\top G_\phi \tilde{W}_a - \tfrac{1}{4} \tilde{W}_a^\top G_\phi W + \tfrac{1}{4} \tilde{W}_a^\top G_\phi \tilde{W}_a \\
&= Q + \tfrac{1}{4} W^\top G_\phi W - \tfrac{1}{2} W^\top G_\phi \tilde{W}_a + \tfrac{1}{4} \tilde{W}_a^\top G_\phi \tilde{W}_a.
\end{aligned}
$$

Combining the preceding terms allows (8.32) to be expressed as

$$\hat{H} = -\omega^\top \tilde{W}_c + W^\top \tfrac{\partial \phi}{\partial x} f + W^\top \tfrac{\partial \phi}{\partial x} F \hat{\theta} - \tfrac{1}{4} W^\top G_\phi W + Q + \tfrac{1}{4} \tilde{W}_a^\top G_\phi \tilde{W}_a. \tag{8.33}$$

We now proceed with a similar analysis for the optimal Hamiltonian (8.5), which can be expressed as

$$H(x, k^*(x), \nabla V^*(x)) = \frac{\partial V^*}{\partial x}(f + F\theta + gk^*) + Q + k^{*\top} Rk^*$$

$$= \frac{\partial V^*}{\partial x}f + \frac{\partial V^*}{\partial x}F\theta + Q - \frac{1}{4}\frac{\partial V^*}{\partial x}G_R\frac{\partial V^*}{\partial x}^\top. \tag{8.34}$$

Using the representation of the value function from (8.8) allows the above to be expressed as

$$H(x, k^*(x), \nabla V^*(x)) = W^\top \frac{\partial \phi}{\partial x}f + \frac{\partial \varepsilon}{\partial x}f + W^\top \frac{\partial \phi}{\partial x}F\theta + \frac{\partial \varepsilon}{\partial x}F\theta + Q - \frac{1}{4}\frac{\partial V^*}{\partial x}G_R\frac{\partial V^*}{\partial x}^\top,$$

where the last term in the above equation can be expanded to obtain

$$\frac{1}{4}\frac{\partial V^*}{\partial x}G_R\frac{\partial V^*}{\partial x}^\top = \frac{1}{4}\frac{\partial V^*}{\partial x}G_R\frac{\partial \phi}{\partial x}^\top W + \frac{1}{4}\frac{\partial V^*}{\partial x}G_R\frac{\partial \phi}{\partial x}^\top\frac{\partial \varepsilon}{\partial x}^\top$$

$$= \frac{1}{4}W^\top G_\phi W + \frac{1}{4}\frac{\partial \varepsilon}{\partial x}G_R\frac{\partial \phi}{\partial x}^\top W + \frac{1}{4}W^\top\frac{\partial \phi}{\partial x}G_R\frac{\partial \varepsilon}{\partial x}^\top + \frac{1}{4}G_\varepsilon$$

$$= \frac{1}{4}W^\top G_\phi W + \frac{1}{2}\frac{\partial \varepsilon}{\partial x}G_R\frac{\partial \phi}{\partial x}^\top W + \frac{1}{4}G_\varepsilon,$$

where $G_\varepsilon(x) := \frac{\partial \varepsilon}{\partial x}(x)G_R(x)\frac{\partial \varepsilon}{\partial x}(x)^\top$. Substituting the above expression back into H then yields

$$H = W^\top \frac{\partial \phi}{\partial x}f + \frac{\partial \varepsilon}{\partial x}f + W^\top \frac{\partial \phi}{\partial x}F\theta + \frac{\partial \varepsilon}{\partial x}F\theta + Q$$

$$- \frac{1}{4}W^\top G_\phi W - \frac{1}{2}\frac{\partial \varepsilon}{\partial x}G_R\frac{\partial \phi}{\partial x}^\top W - \frac{1}{4}G_\varepsilon. \tag{8.35}$$

Recall that the approximate BE is defined as $\hat{\delta} = \hat{H} - H$, hence, subtracting (8.35) away from (8.33) yields the alternate form of the BE

$$\hat{\delta} = -\omega^\top \tilde{W}_c + \frac{1}{4}\tilde{W}_a^\top G_\phi \tilde{W}_a - W^\top \frac{\partial \phi}{\partial x}F\tilde{\theta} + \Delta, \tag{8.36}$$

where

$$\Delta(x) := \frac{1}{2}\frac{\partial \varepsilon}{\partial x}(x)G_R(x)\frac{\partial \phi}{\partial x}^\top W + \frac{1}{4}G_\varepsilon(x) - \frac{\partial \varepsilon}{\partial x}(x)f(x) - \frac{\partial \varepsilon}{\partial x}(x)F(x)\theta. \tag{8.37}$$

Now consider the Lyapunov function candidate

$$V(z, t) = \underbrace{\frac{1}{2}\tilde{W}_c^\top \Gamma^{-1}(t)\tilde{W}_c}_{V_c} + \underbrace{\frac{1}{2}\tilde{W}_a^\top \tilde{W}_a}_{V_a} + V_\theta(\tilde{\theta}, t), \tag{8.38}$$

where V_θ is from Assumption 8.2. Provided the conditions of Lemma 8.1 hold, then V can be bounded for all $(z, t) \in \mathbb{R}^{2l+p} \times \mathbb{R}_{\geq 0}$ as

$$\kappa_1\|z\|^2 \leq V(z, t) \leq \kappa_2\|z\|^2.$$

Computing the derivative of V along the trajectory of z yields

$$\dot{V} = \underbrace{\tilde{W}_c^\top \Gamma^{-1}(t)\dot{\tilde{W}}_c - \frac{1}{2}\tilde{W}_c^\top \Gamma^{-1}(t)\dot{\Gamma}(t)\Gamma^{-1}(t)\tilde{W}_c}_{\dot{V}_c} + \underbrace{\tilde{W}_a^\top \dot{\tilde{W}}_a}_{\dot{V}_a} + \dot{V}_\theta(\tilde{\theta}, t). \tag{8.39}$$

Before substituting in the expressions for the update laws, it will be convenient to express everything in terms of the estimation errors. We begin with $\tilde{W}_c$ and, for ease of exposition, define $\omega_i := \omega(x_i, \hat{W}_a, \hat{\theta})$ $\rho_i := \rho(x_i, \hat{W}_a, \hat{\theta})$, and $\hat{\delta}_i := \hat{\delta}(x_i, \hat{W}_c, \hat{W}_a, \hat{\theta})$, which, after using the alternate form of the approximate BE (8.36), gives us

$$\dot{\tilde{W}}_c = -\dot{\hat{W}}_c$$

$$= \Gamma \frac{\kappa_c}{N} \sum_{i=1}^N \frac{\omega_i}{\rho_i^2}\hat{\delta}_i$$

$$= \Gamma \frac{\kappa_c}{N} \sum_{i=1}^N \frac{\omega_i}{\rho_i^2}\left(-\omega_i^\top \tilde{W}_c + \tfrac{1}{4}\tilde{W}_a^\top G_{\phi,i}\tilde{W}_a - W^\top \frac{\partial \phi}{\partial x}_i F_i\tilde{\theta} + \Delta_i\right)$$

$$= -\Gamma \frac{\kappa_c}{N} \sum_{i=1}^N \frac{\omega_i\omega_i^\top}{\rho_i^2}\tilde{W}_c + \Gamma \frac{\kappa_c}{N} \sum_{i=1}^N \frac{\omega_i}{\rho_i^2}\left(\tfrac{1}{4}\tilde{W}_a^\top G_{\phi,i}\tilde{W}_a - W^\top \frac{\partial \phi}{\partial x}_i F_i\tilde{\theta} + \Delta_i\right),$$

$$\tag{8.40}$$

where $G_{\phi,i} := G_\phi(x_i)$, $\frac{\partial \phi}{\partial x}_i := \frac{\partial \phi}{\partial x}(x_i)$, $F_i := F(x_i)$, and $\Delta_i := \Delta(x_i)$. Using (8.40), to compute $\dot{V}_c$ then yields

$$\dot{V}_c = \tilde{W}_c^\top \Gamma^{-1}\dot{\tilde{W}}_c - \frac{1}{2}\tilde{W}_c^\top \Gamma^{-1}\dot{\Gamma}\Gamma^{-1}\tilde{W}_c$$

$$= \tilde{W}_c^\top\left[-\frac{\kappa_c}{N}\sum_{i=1}^N \frac{\omega_i\omega_i^\top}{\rho_i^2}\tilde{W}_c + \frac{\kappa_c}{N}\sum_{i=1}^N \frac{\omega_i}{\rho_i^2}\left(\tfrac{1}{4}\tilde{W}_a^\top G_{\phi,i}\tilde{W}_a - W^\top \frac{\partial \phi}{\partial x}_i F_i\tilde{\theta} + \Delta_i\right)\right]$$

$$- \tilde{W}_c^\top\left[\frac{\beta}{2}\Gamma^{-1} + \frac{\kappa_c}{2N}\sum_{i=1}^N \frac{\omega_i\omega_i^\top}{\rho_i^2}\right]\tilde{W}_c$$

$$= -\kappa_c\tilde{W}_c^\top\left[\frac{\beta}{2\kappa_c}\Gamma^{-1} + \frac{1}{2N}\sum_{i=1}^N \frac{\omega_i\omega_i^\top}{\rho_i^2}\right]\tilde{W}_c + \tilde{W}_c^\top\left[\frac{\kappa_c}{N}\sum_{i=1}^N \frac{\omega_i}{\rho_i^2}\left(\Delta_i - W^\top \frac{\partial \phi}{\partial x}_i F_i\tilde{\theta}\right)\right]$$

$$+ \tilde{W}_c^\top \frac{\kappa_c}{N}\sum_{i=1}^N \frac{\omega_i\tilde{W}_a^\top G_{\phi,i}\tilde{W}_a}{4\rho_i^2}.$$

$$\tag{8.41}$$

Provided the conditions of Lemma 8.1 hold, then $\dot{V}_c$ can be upper bounded as

$$\dot{V}_c \leq -\kappa_c \left(\frac{\beta}{2\kappa_c \bar{\Gamma}} + \frac{\lambda_c}{2} \right) \|\tilde{W}_c\|^2 + \frac{3\sqrt{3}\kappa_c}{16} \overline{\|\Delta\|_{\mathcal{X}}} \|\tilde{W}_c\|$$

$$+ \frac{3\sqrt{3}\kappa_c}{16} \|W\| \overline{\left\| \frac{\partial\phi}{\partial x} F \right\|_{\mathcal{X}}} \|\tilde{W}_c\| \|\tilde{\theta}\| + \tilde{W}_c^\top \frac{\kappa_c}{N} \sum_{i=1}^N \frac{\omega_i \tilde{W}_a^\top G_{\phi,i} \tilde{W}_a}{4\rho_i^2}$$

$$= -\kappa_c \bar{\lambda} \|\tilde{W}_c\|^2 + \frac{3\sqrt{3}\kappa_c}{16} \overline{\|\Delta\|_{\mathcal{X}}} \|\tilde{W}_c\| + \frac{3\sqrt{3}\kappa_c}{16} \|W\| \overline{\left\| \frac{\partial\phi}{\partial x} F \right\|_{\mathcal{X}}} \|\tilde{W}_c\| \|\tilde{\theta}\|$$

$$+ \tilde{W}_c^\top \frac{\kappa_c}{N} \sum_{i=1}^N \frac{\omega_i \tilde{W}_a^\top G_{\phi,i} \tilde{W}_a}{4\rho_i^2}, \tag{8.42}$$

where the bound follows from the fact that for any $\omega \in \mathbb{R}^l$

$$\left\| \frac{\omega}{(1 + \omega^\top \omega)^2} \right\| \leq \frac{3\sqrt{3}}{16}.$$

We now proceed to analyze $\tilde{W}_a$:

$$\dot{\tilde{W}}_a = -\dot{\hat{W}}_a$$

$$= \kappa_{a_1}(\hat{W}_a - \hat{W}_c) + \kappa_{a_2}\hat{W}_a - \frac{\kappa_c}{N} \sum_{i=1}^N \frac{G_\phi(x_i)\hat{W}_a \omega_i^\top}{4\rho_i^2} \hat{W}_c$$

$$= \kappa_{a_1}(W - \tilde{W}_a - W + \tilde{W}_c) + \kappa_{a_2}W - \kappa_{a_2}\tilde{W}_a - \frac{\kappa_c}{N} \sum_{i=1}^N \frac{G_\phi(x_i)\hat{W}_a \omega_i^\top}{4\rho_i^2} \hat{W}_c$$

$$= -(\kappa_{a_1} + \kappa_{a_2})\tilde{W}_a + \kappa_{a_1}\tilde{W}_c + \kappa_{a_2}W - \frac{\kappa_c}{N} \sum_{i=1}^N \frac{G_\phi(x_i)\hat{W}_a \omega_i^\top}{4\rho_i^2} \hat{W}_c.$$

The last term in the preceding equation can be expanded as

$$\frac{G_\phi \hat{W}_a \omega^\top}{4\rho^2} \hat{W}_c = \frac{G_\phi \hat{W}_a \omega^\top}{4\rho^2} W - \frac{G_\phi \hat{W}_a \omega^\top}{4\rho^2} \tilde{W}_c$$

$$= \frac{G_\phi W \omega^\top}{4\rho^2} W - \frac{G_\phi \tilde{W}_a \omega^\top}{4\rho^2} W - \frac{G_\phi W \omega^\top}{4\rho^2} \tilde{W}_c + \frac{G_\phi \tilde{W}_a \omega^\top}{4\rho^2} \tilde{W}_c,$$

which implies that

$$\dot{\tilde{W}}_a = -(\kappa_{a_1} + \kappa_{a_2})\tilde{W}_a + \kappa_{a_1}\tilde{W}_c + \kappa_{a_2}W - \frac{\kappa_c}{N} \sum_{i=1}^N \frac{G_{\phi,i} W \omega_i^\top}{4\rho_i^2} W$$

$$+ \frac{\kappa_c}{N} \sum_{i=1}^N \frac{G_{\phi,i} \tilde{W}_a \omega_i^\top}{4\rho_i^2} W + \frac{\kappa_c}{N} \sum_{i=1}^N \frac{G_{\phi,i} W \omega_i^\top}{4\rho_i^2} \tilde{W}_c - \frac{\kappa_c}{N} \sum_{i=1}^N \frac{G_{\phi,i} \tilde{W}_a \omega_i^\top}{4\rho_i^2} \tilde{W}_c,$$

Using the above equation to compute $\dot{V}_a$ yields

$$\dot{V}_a = \tilde{W}_a^\top \dot{\tilde{W}}_a$$

$$= -(\kappa_{a_1} + \kappa_{a_2})\|\tilde{W}_a\|^2 + \kappa_{a_1}\tilde{W}_a^\top \tilde{W}_c + \kappa_{a_2}\tilde{W}_a^\top W - \tilde{W}_a^\top \frac{\kappa_c}{N} \sum_{i=1}^{N} \frac{G_{\phi,i} W \omega_i^\top}{4\rho_i^2} W$$

$$+ \tilde{W}_a^\top \frac{\kappa_c}{N} \sum_{i=1}^{N} \left(\frac{G_{\phi,i}\tilde{W}_a \omega_i^\top}{4\rho_i^2} W + \frac{G_{\phi,i} W \omega_i^\top}{4\rho_i^2} \tilde{W}_c - \frac{G_{\phi,i}\tilde{W}_a \omega_i^\top}{4\rho_i^2} \tilde{W}_c \right).$$

Upper bounding the above expression then yields

$$\dot{V}_a \leq -(\kappa_{a_1} + \kappa_{a_2})\|\tilde{W}_a\|^2 + \kappa_{a_1}\|\tilde{W}_a\|\|\tilde{W}_c\| + \kappa_{a_2}\|W\|\|\tilde{W}_a\|$$

$$+ \frac{3\sqrt{3}\kappa_c}{64}\left(\|W\|^2\overline{\|G_\phi\|}_\chi\right)\|\tilde{W}_a\| + \frac{3\sqrt{3}\kappa_c}{64}\left(\|W\|\overline{\|G_\phi\|}_\chi\right)\|\tilde{W}_a\|^2$$

$$+ \frac{3\sqrt{3}\kappa_c}{64}\left(\|W\|\overline{\|G_\phi\|}_\chi\right)\|\tilde{W}_a\|\|\tilde{W}_c\| - \tilde{W}_a^\top \frac{\kappa_c}{N} \sum_{i=1}^{N} \frac{G_{\phi,i}\tilde{W}_a \omega_i^\top}{4\rho_i^2} \tilde{W}_c.$$

After grouping similar terms, the above bound can be expressed as

$$\dot{V}_a \leq \left(\frac{3\sqrt{3}\kappa_c}{64}\|W\|\overline{\|G_\phi\|}_\chi - (\kappa_{a_1} + \kappa_{a_2}) \right) \|\tilde{W}_a\|^2$$

$$+ \left(\kappa_{a_2}\|W\| + \frac{3\sqrt{3}\kappa_c}{64}\|W\|^2\overline{\|G_\phi\|}_\chi \right) \|\tilde{W}_a\| \tag{8.43}$$

$$+ \left(\kappa_{a_1} + \frac{3\sqrt{3}\kappa_c}{64}\|W\|\overline{\|G_\phi\|}_\chi \right) \|\tilde{W}_a\|\|\tilde{W}_c\| - \tilde{W}_a^\top \frac{\kappa_c}{N} \sum_{i=1}^{N} \frac{G_{\phi,i}\tilde{W}_a \omega_i^\top}{4\rho_i^2} \tilde{W}_c.$$

Now, adding $\dot{V}_c$ and $\dot{V}_a$, taking upper bounds using (8.42) and (8.43), and recognizing that the last term in (8.42) and (8.43) cancel out yields

$$\dot{V}_c + \dot{V}_a \leq -\kappa_c \bar{\lambda}\|\tilde{W}_c\|^2 - \left((\kappa_{a_1} + \kappa_{a_2}) - \frac{3\sqrt{3}\kappa_c}{64}\|W\|\overline{\|G_\phi\|}_\chi \right) \|\tilde{W}_a\|^2$$

$$+ \frac{3\sqrt{3}\kappa_c}{16}\overline{\|\Delta\|}_\chi\|\tilde{W}_c\| + \left(\kappa_{a_2}\|W\| + \frac{3\sqrt{3}\kappa_c}{64}\|W\|^2\overline{\|G_\phi\|}_\chi \right) \|\tilde{W}_a\|$$

$$+ \left(\kappa_{a_1} + \frac{3\sqrt{3}\kappa_c}{64}\|W\|\overline{\|G_\phi\|}_\chi \right) \|\tilde{W}_a\|\|\tilde{W}_c\| + \left(\frac{3\sqrt{3}\kappa_c}{16}\|W\|\overline{\left\|\frac{\partial\phi}{\partial x}F\right\|}_\chi \right) \|\tilde{W}_c\|\|\tilde{\theta}\|.$$

$$\tag{8.44}$$

Defining

$$\iota_c := \frac{3\sqrt{3}\kappa_c}{16}\overline{\|\Delta\|}_\chi$$

$$\iota_a := \kappa_{a_2}\|W\| + \frac{3\sqrt{3}\kappa_c}{64}\|W\|^2\overline{\|G_\phi\|}_\chi$$

$$\upsilon_{ac} := \kappa_{a_1} + \frac{3\sqrt{3}\kappa_c}{64}\|W\|\overline{\|G_\phi\|}_\chi$$

$$\upsilon_{c\theta} := \frac{3\sqrt{3}\kappa_c}{16}\|W\|\overline{\left\|\frac{\partial\phi}{\partial x}F\right\|}_\chi$$

$$\upsilon_a := \frac{3\sqrt{3}\kappa_c}{64}\|W\|\overline{\|G_\phi\|}_\chi,$$

allows the bound in (8.44) to be compactly represented as

$$\dot{V}_c + \dot{V}_a \leq -\kappa_c\bar{\lambda}\|\tilde{W}_c\|^2 - \left((\kappa_{a_1} + \kappa_{a_2}) - \upsilon_a\right)\|\tilde{W}_a\|^2 + \iota_c\|\tilde{W}_c\| + \iota_a\|\tilde{W}_a\|$$
$$+ \upsilon_{ac}\|\tilde{W}_a\|\|\tilde{W}_c\| + \upsilon_{c\theta}\|\tilde{W}_c\|\|\tilde{\theta}\|.$$

Combining the above bound with that on $\dot{V}_\theta$ from Assumption 8.2 allows $\dot{V}$ to be bounded as

$$\dot{V} \leq -\kappa_c\bar{\lambda}\|\tilde{W}_c\|^2 - (\kappa_{a_1} + \kappa_{a_2})\|\tilde{W}_a\|^2 + \upsilon_a\|\tilde{W}_a\|^2 - c_3\|\tilde{\theta}\|^2$$
$$+ \iota_c\|\tilde{W}_c\| + \iota_a\|\tilde{W}_a\| + \upsilon_{ac}\|\tilde{W}_a\|\|\tilde{W}_c\| + \upsilon_{c\theta}\|\tilde{W}_c\|\|\tilde{\theta}\|. \tag{8.45}$$

The objective is now to complete squares in the above relation to show that $\dot{V} < 0$ for sufficiently large weight estimation errors. To this end, observe that

$$-\frac{\kappa_c\bar{\lambda}}{2}\|\tilde{W}_c\|^2 + \iota_c\|\tilde{W}_c\| \leq \frac{\iota_c^2}{2\kappa_c\bar{\lambda}},$$

$$-\frac{1}{2}(\kappa_{a_1} + \kappa_{a_2})\|\tilde{W}_a\|^2 + \iota_a\|\tilde{W}_a\| \leq \frac{\iota_a^2}{2(\kappa_{a_1} + \kappa_{a_2})},$$

which allows $\dot{V}$ to be further bounded as

$$\dot{V} \leq -\frac{\kappa_c\bar{\lambda}}{2}\|\tilde{W}_c\|^2 - \frac{\kappa_{a_1} + \kappa_{a_2}}{2}\|\tilde{W}_a\|^2 + \upsilon_a\|\tilde{W}_a\|^2 - c_3\|\tilde{\theta}\|^2$$
$$+ \upsilon_{ac}\|\tilde{W}_a\|\|\tilde{W}_c\| + \upsilon_{c\theta}\|\tilde{W}_c\|\|\tilde{\theta}\| + \iota,$$

where

$$\iota := \frac{\iota_c^2}{2\kappa_c\bar{\lambda}} + \frac{\iota_a^2}{2(\kappa_{a_1} + \kappa_{a_2})}.$$

Partitioning terms in the preceding bound allows the bound on $\dot{V}$ to be expressed as

$$\dot{V} \leq -\frac{\kappa_c \bar{\lambda}}{4}\|\tilde{W}_c\|^2 - \frac{\kappa_{a_1} + \kappa_{a_2}}{4}\|\tilde{W}_a\|^2 - \frac{c_3}{2}\|\tilde{\theta}\|^2 + \iota$$

$$- \begin{bmatrix} \|\tilde{W}_c\| & \|\tilde{W}_a\| & \|\tilde{\theta}\| \end{bmatrix} \underbrace{\begin{bmatrix} \frac{\kappa_c \bar{\lambda}}{4} & -\frac{v_{ac}}{2} & -\frac{v_{c\theta}}{2} \\ -\frac{v_{ac}}{2} & \frac{\kappa_{a_1}+\kappa_{a_2}}{4} - v_a & 0 \\ -\frac{v_{c\theta}}{2} & 0 & \frac{c_3}{2} \end{bmatrix}}_{M} \begin{bmatrix} \|\tilde{W}_c\| \\ \|\tilde{W}_a\| \\ \|\tilde{\theta}\| \end{bmatrix}. \tag{8.46}$$

Provided that $\lambda_{\min}(M) > 0$, the above can be further bounded as

$$\dot{V} \leq -\frac{\kappa_c \bar{\lambda}}{4}\|\tilde{W}_c\|^2 - \frac{\kappa_{a_1} + \kappa_{a_2}}{4}\|\tilde{W}_a\|^2 - \frac{c_3}{2}\|\tilde{\theta}\|^2 + \iota$$

$$\leq -\kappa_3\|z\|^2 + \iota \tag{8.47}$$

$$\leq -\frac{\kappa_3}{\kappa_2}V + \iota.$$

Invoking the Comparison Lemma (Lemma 2.1) implies that for all $t \in \mathbb{R}_{\geq 0}$

$$V(z(t), t) \leq V(z(0), 0)e^{-\frac{\kappa_3}{\kappa_2}t} + (1 - e^{-\frac{\kappa_3}{\kappa_2}t})\frac{\iota}{\kappa_3} \tag{8.48}$$

which, after combining with the bounds on V, implies that for all $t \in \mathbb{R}_{\geq 0}$

$$\|z(t)\| \leq \sqrt{\frac{\kappa_2}{\kappa_1}\|z(0)\|^2 e^{-\frac{\kappa_3}{\kappa_2}t} + (1 - e^{-\frac{\kappa_3}{\kappa_2}t})\frac{\iota}{\kappa_1\kappa_3}}, \tag{8.49}$$

which is exactly the bound from (8.30), as desired. $\square$

Having established convergence of the estimated weights to a neighborhood of their ideal values, we now establish stability of the closed-loop system under the MBRL-based policy from (8.10), which is the learned policy used to generate data and *not* the safe policy (8.20) ultimately applied to the system.[2] The stability analysis is facilitated by the Lyapunov function candidate

$$V_L(y, t) := V^*(x) + V(z, t), \tag{8.50}$$

where $y := [x^\top \ z^\top]^\top$ is a composite state vector with $z \in \mathbb{R}^{2l+p}$ defined as in Theorem 8.2 and V is the Lyapunov function candidate from the proof of Theorem 8.2. We recall that since V_L is positive definite, there exist of $\alpha_1, \alpha_2 \in \mathcal{K}$ satisfying

$$\alpha_1(\|y\|) \leq V_L(y, t) \leq \alpha_2(\|y\|). \tag{8.51}$$

Before stating the result, we require one more Lyapunov theorem that we have not yet introduced.

---

[2] Results on stability using the policy from (8.20) are postponed until the next chapter.

Theorem 8.3 (Uniformly ultimately bounded) *Let* $f : \mathbb{R}^n \times \mathbb{R}_{\geq 0} \to \mathbb{R}^n$ *be a vector field, locally Lipschitz in its first argument and piecewise continuous in its second argument, that induces the non-autonomous dynamical system* $\dot{x} = f(x, t)$ *defined on some domain* $X \subset \mathbb{R}^n$. *Let* $V : \mathbb{R}^n \times \mathbb{R}_{\geq 0} \to \mathbb{R}_{\geq 0}$ *be a continuously differentiable function satisfying*

$$\alpha_1(\|x\|) \leq V(x, t) \leq \alpha_2(\|x\|), \quad \forall(x, t) \in X \times \mathbb{R}_{\geq 0}, \tag{8.52a}$$

$$\dot{V}(x, t) \leq -W(x), \quad \forall \|x\| \geq \mu > 0, \ \forall t \in \mathbb{R}_{\geq 0}, \tag{8.52b}$$

where $\alpha_1, \alpha_2 \in \mathcal{K}$ *and* $W : \mathbb{R}^n \to \mathbb{R}_{\geq 0}$ *is continuous and positive definite. Let* $\bar{B}_r(0) \subset X$ *be a closed ball of radius* $r \in \mathbb{R}_{>0}$ *contained in* X *centered at the origin such that* $\mu < \alpha_2^{-1}(\alpha_1(r))$. *Then, for any initial condition* $x_0 \in X$ *satisfying* $\|x_0\| \leq \alpha_2^{-1}(\alpha_1(r))$ *there exists a* $\beta \in \mathcal{KL}$ *and a time* $T \in \mathbb{R}_{\geq 0}$ *such that the trajectory* $t \mapsto x(t)$ *with* $x(0) = x_0$ *satisfies*

$$\|x(t)\| \leq \beta(\|x(0)\|, t), \quad \forall t \in [0, T], \tag{8.53a}$$

$$\|x(t)\| \leq \alpha_1^{-1}(\alpha_2(\mu)), \quad \forall t \in [T, \infty). \tag{8.53b}$$

The above theorem allows one to establish asymptotic convergence of trajectories to a ball about the origin and is very similar to the notion of input-to-state stability used in Chap. 6. A trajectory satisfying (8.53) is said to be *uniformly ultimately bounded*. We now have all the tools in place to provide conditions under which the learning-based controller guarantees stability.

Theorem 8.4 *Consider system (4.1) under the influence of the learning-based policy from (8.10). Let* $y := [x^\top \ \tilde{W}_c^\top \ \tilde{W}_a^\top \ \tilde{\theta}^\top]^\top \in \mathbb{R}^{n+2l+p}$ *be a composite state vector and suppose the estimated weights and parameters are updated according to (8.23), (8.24), (8.25), and (5.25). Let* $\bar{\mathcal{B}}_r(0) \subset X \times \mathbb{R}^{2l+p}$ *be a closed ball of radius* $r \in \mathbb{R}_{>0}$ *contained in* $X \times \mathbb{R}^{2l+p}$. *Provided the conditions of Theorem 8.2 hold and*

$$\mu := \alpha_3^{-1}(2\nu) < \alpha_2^{-1}(\alpha_1(r)),$$

where

$$\nu := \frac{\iota_c^2}{2\kappa_c \bar{\lambda}} + \frac{\iota_{a2}^2}{2(\kappa_{a_1} + \kappa_{a_2})} + \frac{1}{4}\overline{\|G_\varepsilon(x)\|}_X \tag{8.54}$$

$$\iota_{a2} := \iota_a + \frac{1}{2}\left\|W^\top G_\phi(x) + \nabla\varepsilon(x)G_R(x)\frac{\partial\phi}{\partial x}(x)^\top\right\|_X,$$

and $\alpha_3 \in \mathcal{K}$ *satisfies*

$$\alpha_3(\|y\|) \leq Q(x) + \frac{\bar{\lambda}\kappa_c}{4}\|\tilde{W}_c\|^2 + \frac{\kappa_{a_1} + \kappa_{a_2}}{4}\|\tilde{W}_a\|^2 + \frac{c_3}{2}\|\tilde{\theta}\|^2, \tag{8.55}$$

then any trajectory $t \mapsto y(t)$ *with an initial condition such that*

$$y(0) \le \alpha_2^{-1}(\alpha_1(r)),$$

satisfies

$$\limsup_{t \to \infty} \|y(t)\| \le \alpha_1^{-1}(\alpha_2(\mu)). \tag{8.56}$$

Proof The proof is facilitated by the Lyapunov function candidate V_L from (8.50) composed the optimal value function and the Lyapunov function used in the proof of Theorem 8.2. We begin by analyzing V^* whose Lie derivative along the closed-loop dynamics (4.1) equipped with the controller from (8.10) is

$$
\begin{aligned}
\dot{V}^* &= \frac{\partial V^*}{\partial x}(f + F\theta + g\hat{k}) \\
&= \frac{\partial V^*}{\partial x}(f + F\theta) - \frac{1}{2}\frac{\partial V^*}{\partial x}G_R\frac{\partial \phi}{\partial x}^\top \hat{W}_a \\
&= \frac{\partial V^*}{\partial x}(f + F\theta) - \frac{1}{2}\frac{\partial V^*}{\partial x}G_R\frac{\partial \phi}{\partial x}^\top W + \frac{1}{2}\frac{\partial V^*}{\partial x}G_R\frac{\partial \phi}{\partial x}^\top \tilde{W}_a \\
&= \frac{\partial V^*}{\partial x}(f + F\theta) - \frac{1}{2}W^\top G_\phi W - \frac{1}{2}\frac{\partial \varepsilon}{\partial x}G_R\frac{\partial \phi}{\partial x}^\top W + \frac{1}{2}W^\top G_\phi \tilde{W}_a + \frac{1}{2}\frac{\partial \varepsilon}{\partial x}G_R\frac{\partial \phi}{\partial x}^\top \tilde{W}_a.
\end{aligned}
\tag{8.57}
$$

Using the alternate form of the optimal Hamiltonian from (8.35), we have that

$$\frac{\partial V^*}{\partial x}(f + F\theta) = -Q + \frac{1}{4}W^\top G_\phi W + \frac{1}{2}\frac{\partial \varepsilon}{\partial x}G_R\frac{\partial \phi}{\partial x}^\top W + \frac{1}{4}G_\varepsilon,$$

which, after substituting into the preceding expression and upper bounding yields

$$
\begin{aligned}
\dot{V}^* &= -Q - \frac{1}{4}W^\top G_\phi W + \frac{1}{4}G_\varepsilon + \frac{1}{2}(W^\top G_\phi + \frac{\partial \varepsilon}{\partial x}G_R\frac{\partial \phi}{\partial x}^\top)\tilde{W}_a \\
&\le -Q + \frac{1}{4}\overline{\|G_\varepsilon\|}_\chi + \frac{1}{2}\left\| W^\top G_\phi + \frac{\partial \varepsilon}{\partial x}G_R\frac{\partial \phi}{\partial x}^\top \right\|_\chi \|\tilde{W}_a\|.
\end{aligned}
\tag{8.58}
$$

Now, taking the Lie derivative of V_L along the composite system trajectory and then upper bounding using the bounds on $\dot{V}$ and $\dot{V}^*$ from (8.45) and (8.58), respectively, yields

$$
\begin{aligned}
\dot{V}_L &= \dot{V}^* + \dot{V} \\
&\le -Q + \frac{1}{4}\overline{\|G_\varepsilon\|}_\chi + \frac{1}{2}\left\| W^\top G_\phi + \frac{\partial \varepsilon}{\partial x}G_R\frac{\partial \phi}{\partial x}^\top \right\|_\chi \|\tilde{W}_a\| \\
&\quad - \kappa_c \bar{\lambda}\|\tilde{W}_c\|^2 - (\kappa_{a_1} + \kappa_{a_2})\|\tilde{W}_a\|^2 + \upsilon_a\|\tilde{W}_a\|^2 - c_3\|\tilde{\theta}\|^2 \\
&\quad + \iota_c\|\tilde{W}_c\| + \iota_a\|\tilde{W}_a\| + \upsilon_{ac}\|\tilde{W}_a\|\|\tilde{W}_c\| + \upsilon_{c\theta}\|\tilde{W}_c\|\|\tilde{\theta}\| \\
&= -Q - \kappa_c \bar{\lambda}\|\tilde{W}_c\|^2 - (\kappa_{a_1} + \kappa_{a_2})\|\tilde{W}_a\|^2 + \upsilon_a\|\tilde{W}_a\|^2 - c_3\|\tilde{\theta}\|^2 \\
&\quad + \iota_c\|\tilde{W}_c\| + \iota_{a2}\|\tilde{W}_a\| + \upsilon_{ac}\|\tilde{W}_a\|\|\tilde{W}_c\| + \upsilon_{c\theta}\|\tilde{W}_c\|\|\tilde{\theta}\| + \frac{1}{4}\overline{\|G_\varepsilon\|}_\chi,
\end{aligned}
\tag{8.59}
$$

where

$$\iota_{a2} := \iota_a + \frac{1}{2}\left\| W^\top G_\phi + \frac{\partial \varepsilon}{\partial x}G_R\frac{\partial \phi}{\partial x}^\top \right\|_\chi.$$

From this point the proof follows similar steps to those of Theorem 8.2: after separating terms, completing squares, and further bounding we obtain

$$\dot{V}_L \leq - Q - \frac{\kappa_c \bar{\lambda}}{2} \|\tilde{W}_c\|^2 - \frac{\kappa_{a_1} + \kappa_{a_2}}{2} \|\tilde{W}_a\|^2 + v_a \|\tilde{W}_a\|^2 - c_3 \|\tilde{\theta}\|^2$$
$$+ v_{ac} \|\tilde{W}_a\| \|\tilde{W}_c\| + v_{c\theta} \|\tilde{W}_c\| \|\tilde{\theta}\| + v,$$

where

$$v := \frac{\iota_c^2}{2\kappa_c \bar{\lambda}} + \frac{\iota_{a2}^2}{2(\kappa_{a_1} + \kappa_{a_2})} + \frac{1}{4} \overline{\|G_\varepsilon(x)\|_X}$$

Partitioning terms in the preceding bound allows the bound on $\dot{V}_L$ to be expressed as

$$\dot{V}_L \leq - Q - \frac{\kappa_c \bar{\lambda}}{4} \|\tilde{W}_c\|^2 - \frac{\kappa_{a_1} + \kappa_{a_2}}{4} \|\tilde{W}_a\|^2 - \frac{c_3}{2} \|\tilde{\theta}\|^2 + v$$
$$- \begin{bmatrix} \|\tilde{W}_c\| & \|\tilde{W}_a\| & \|\tilde{\theta}\| \end{bmatrix} \underbrace{\begin{bmatrix} \frac{\kappa_c \bar{\lambda}}{4} & -\frac{v_{ac}}{2} & -\frac{v_{c\theta}}{2} \\ -\frac{v_{ac}}{2} & \frac{\kappa_{a_1} + \kappa_{a_2}}{4} - v_a & 0 \\ -\frac{v_{c\theta}}{2} & 0 & \frac{c_3}{2} \end{bmatrix}}_{M} \begin{bmatrix} \|\tilde{W}_c\| \\ \|\tilde{W}_a\| \\ \|\tilde{\theta}\| \end{bmatrix}. \quad (8.60)$$

Provided the conditions of Theorem 8.2 hold, then M is positive definite and $\dot{V}_L$ can be further bounded as

$$\dot{V}_L \leq - Q - \frac{\kappa_c \bar{\lambda}}{4} \|\tilde{W}_c\|^2 - \frac{\kappa_{a_1} + \kappa_{a_2}}{4} \|\tilde{W}_a\|^2 - \frac{c_3}{2} \|\tilde{\theta}\|^2 + v \quad (8.61)$$

$$\leq - \alpha_3(\|y\|) + v,$$

where $\alpha_3 \in \mathcal{K}$ is any class $\mathcal{K}$ function satisfying

$$\alpha_3(\|y\|) \leq Q(x) + \frac{\kappa_c \bar{\lambda}}{4} \|\tilde{W}_c\|^2 + \frac{\kappa_{a_1} + \kappa_{a_2}}{4} \|\tilde{W}_a\|^2 + \frac{c_3}{2} \|\tilde{\theta}\|^2.$$

The bound in (8.61) implies that

$$\dot{V}_L \leq -\tfrac{1}{2} \alpha_3(\|y\|), \quad \forall \|y\| \geq \underbrace{\alpha_3^{-1}(2v)}_{\mu}. \quad (8.62)$$

Hence, provided that $\mu < \alpha_2^{-1}(\alpha_1(r))$ then Theorem 8.3 implies the existence of a $\beta \in \mathcal{KL}$ and a time $T \in \mathbb{R}_{\geq 0}$ such that for any initial condition $y_0 := y(0)$ satisfying $\|y_0\| \leq \alpha_2^{-1}(\alpha_1(r))$ the resulting solution $t \mapsto y(t)$ satisfies

$$\|y(t)\| \leq \beta(\|y_0\|, t), \qquad \forall t \in [0, T],$$
$$\|y(t)\| \leq \alpha_1^{-1}(\alpha_2(\mu)), \qquad \forall t \in [T, \infty),$$

which implies

$$\limsup_{t \to \infty} \|y(t)\| \le \alpha_1^{-1}(\alpha_2(\mu)),$$

as desired.   □

We again remark that the above theorem does not establish stability under the safe RL policy from (8.20); rather, it establishes stability under the nominal (potentially unsafe) RL policy from (8.10). We note that, similar to Theorem 8.2, verifying the conditions of the preceding theorem are challenging, and, in practice, one must generally resort to trial-and-error tuning of the hyperparameters associated with the algorithm (e.g., number of basis functions, learning gains, number of sampling points, etc.) to produce a stabilizing control policy. Despite this, note that the learning-based policy from (8.20) is safe by definition as the RaCBF conditions are independent of any conditions associated with the RL approach.

8.4 Numerical Examples

Example 8.1 Our first example reconsiders the two-dimensional nonlinear system and safe set from Example 7.2. The system dynamics are in the form of (4.1) and are given by

$$\begin{bmatrix} \dot{x}_1 \\ \dot{x}_2 \end{bmatrix} = \underbrace{\begin{bmatrix} 0 \\ 0 \end{bmatrix}}_{f(x)} + \underbrace{\begin{bmatrix} x_1 & x_2 & 0 \\ 0 & 0 & x_1^3 \end{bmatrix}}_{F(x)} \underbrace{\begin{bmatrix} \theta_1 \\ \theta_2 \\ \theta_3 \end{bmatrix}}_{\theta} + \underbrace{\begin{bmatrix} 0 \\ x_2 \end{bmatrix}}_{g(x)} u,$$

where the uncertain parameters are set to $\theta = [-0.6 \ -1 \ 1]^\top$. The main objective is to drive the system to the origin while remaining in a safe set $C \subset \mathbb{R}^2$ defined as the zero superlevel set of

$$h(x) = 1 - x_1 - x_2^2,$$

and minimizing the infinite-horizon cost (8.1) with a running cost (8.2) defined as

$$\ell(x, u) = \tfrac{1}{2}\|x\|^2 + u^2.$$

The uncertain parameters are learned online using the update law from (5.25) with $\gamma_c = 1$ and a history stack with $M = 20$ entries, where the integration window is chosen as $\Delta t = 0.5$. To guarantee safety using the RaCBF-QP (8.20) we compute bounds on the initial estimation error by assuming that the parameters belong to the set

$$\theta \in [-1, 0] \times [-1.5. \ -0.5] \times [0.5, 1.5] \subset \mathbb{R}^3,$$

and choose the extended class $\mathcal{K}_\infty$ function as $\alpha(s) = 10s^3$. The value function is approximated using the quadratic basis

$$\phi(x) = \begin{bmatrix} x_1^2 \\ x_1 x_2 \\ x_2^2, \end{bmatrix}$$

and the sample points used to evaluate the BE are chosen as the vertices of a uniform grid over $[-4, 1] \times [-2, 2] \subset \mathbb{R}^2$ for a total of $N = 99$ samples. These samples are used to update the weight estimates using the update laws from (8.23), (8.24), and (8.26) with $\kappa_c = 1$, $\beta = 1$, $\kappa_a = 1$.

To demonstrate the efficacy of the approach, we simulate the system with and without shielding the learned policy with the RaCBF-QP (8.20). For each simulation, the initial state is taken as $x(0) = [-4 \ 1]^\top$, the initial value function and policy weights are drawn from a uniform distribution[3] between 0 and 1, and the initial model parameter estimate is taken as $\hat{\theta}(0) = [-0.5 \ -0.6, \ 0.7]^\top$. The results of the simulations are provided in Figs. 8.1 and 8.2.

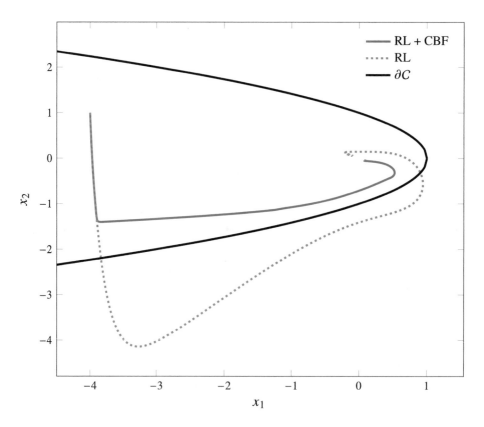

Fig. 8.1 Trajectories of the nonlinear system under the RL policy (8.10) with (solid blue curve) and without (dotted green curve) intervention from the RaCBF-QP (8.20). The black curve denotes the boundary of the safe set

---

[3] The same random weights are used for both simulations.

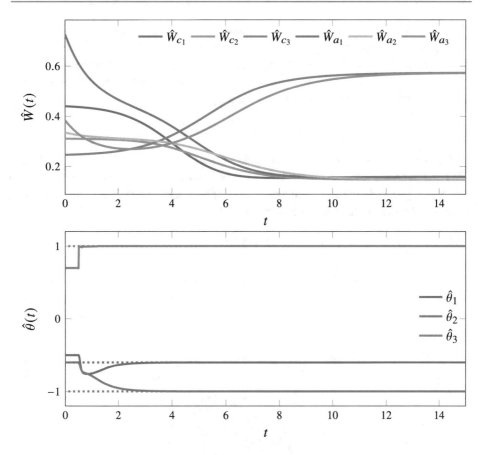

Fig. 8.2 Evolution of the value and policy weights (top) as well as the parameter estimates (bottom) for the nonlinear system under the RaCBF-QP based safety filter from (8.20), which corresponds to the state trajectory given by the blue curve in Fig. 8.1. In the bottom plot, the dotted lines of corresponding color indicate the true values of the model parameters and the solid curve indicate the parameter estimates over time, which converge to their true value in just under 4 s

In particular, Fig. 8.1 illustrates the trajectories of the system under the controller with and without shielding, whereas Fig. 8.2 illustrates the evolution of the value and policy weights, as well as the estimated model parameters. As shown in Fig. 8.1, shielding the RL policy keeps the system safe, whereas the system leaves the safe set under the RL policy without any shielding. Moreover, similar to previous chapters, the adaptive CBF safety filter gradually allows the system to come closer to the boundary of the safe set as the uncertain model parameters are identified (see Fig. 8.2). As the value function for this problem is unknown, the accuracy of the value function and policy weights is challenging to quantify; however, in each of the simulations the weights converge on a policy that stabilizes the system to the origin as predicted by Theorem 8.2.

Example 8.2 We now examine a simple scenario to demonstrate some interesting properties of the proposed online learning approach. We reconsider the robot motion planning example from Example 3.2 in which a mobile robot is modeled as a single integrator $\dot{x} = u$ and the goal is to drive the robot from its initial position to the origin while avoiding an obstacle. The safety objective can be considered by defining

$$h(x) = \|x - x_o\|^2 - r_o^2,$$

where $x_o \in \mathbb{R}^2$ is the location of the obstacle's center and $r_o \in \mathbb{R}_{>0}$ is its radius, which is used to construct a safe set $C \subset \mathbb{R}^2$ as in (3.3) as well as a CBF with $\alpha(s) = s^3$. To obtain a stabilizing control policy we associate to the single integrator an infinite-horizon optimal control problem as in (8.1) with a running cost of

$$\ell(x, u) = \|x\|^2 + \|u\|^2.$$

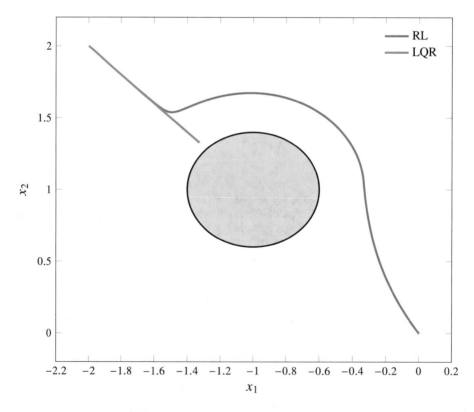

Fig. 8.3 Trajectories of the single integrator under the safe RL policy (blue curve) and safe LQR policy (green curve). The gray disk denotes the obstacle

As the system is linear and the cost is quadratic, the HJB (8.4) reduces to the algebraic Ricatti equation, which can be solved using standard numerical tools to obtain the optimal value function as

$$V^*(x) = x_1^2 + x_2^2.$$

In principle, one could use the linear quadratic regulator (LQR) policy induced by the above value function as the nominal controller in the standard CBF-QP (3.12) to solve this problem; however, as demonstrated in the subsequent numerical results, such an approach presents certain limitations. To compare the LQR solution with the RL solution presented in this chapter, we approximate the value function using the same basis in the preceding example—this implies that the optimal value function weights are $W = [1\ 0\ 1]^\top$ and that there exists no function reconstruction error ε. To learn the value function online, we evaluate the approximate BE (8.15) at $N = 25$ points sampled from a multivariate Gaussian distribution using the current state as the mean and a covariance matrix of $0.1 I_{2\times2}$ at every instant in time. The sampled BE is then used to update the value function weights using the update law in (8.23) and (8.24) with $\kappa_c = 1$ and $\beta = 0.001$. The control policy weights are updated once again using the simple projection update law from Remark 8.2 with $\kappa_a = 1$. Since the dynamics for this example are trivial, no parameter identification is performed.

The results of the simulations comparing the performance of the RL-based policy and the LQR policy, both of which are filtered through the CBF-QP (3.12), are provided in Figs. 8.3 and 8.4. The resulting system trajectories are illustrated in Fig. 8.3, where the trajectory under the RL policy navigates around the obstacle and converges to the origin, whereas the trajectory under the LQR policy gets stuck behind the obstacle. Note that the LQR controller filtered through the CBF-QP is a continuous *time-invariant* feedback controller, and, as discussed in Chap. 3, there are fundamental limitations to the behavior that can be produced by continuous time-invariant vector fields. On the other hand, the RL controller is continuous but is also a dynamic feedback controller as it explicitly depends on the control policy weights $\hat{W}_a$, which are updated based upon data observed online. As seen in Fig. 8.4 (top), the trajectories under both policies quickly approach the obstacle and initially fail to make progress towards the goal. However, unlike the static LQR policy, the weights of the RL controller continue to evolve, as shown in Fig. 8.4 (bottom) and eventually converge to a policy that navigates the robot around the obstacle and to the goal. Furthermore, by the end of the simulation the weights have converged extremely close to their optimal values in line with the results of Theorem 8.2.

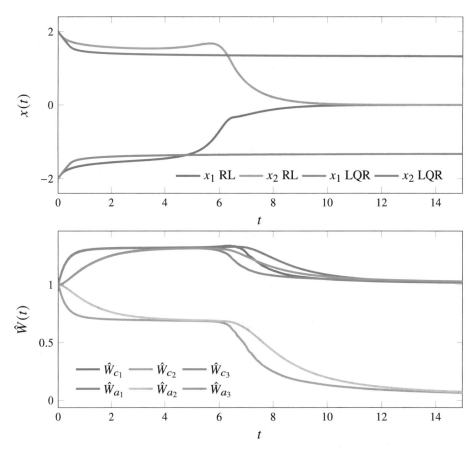

Fig. 8.4 Top: Evolution of the system states under the safe RL and safe LQR policy. Bottom: Evolution of the estimated value function and control policy weights of the the RL policy

8.5   Notes

In this chapter, we presented a safe online reinforcement learning (RL) framework that allows for simultaneously learning the value function and dynamics of a nonlinear system in real-time while satisfying safety constraints. The roots of the online RL method developed in this chapter can be traced back to the work of Vamvoudakis and Lewis in [1]. This work proposed an "actor-critic" structure in which two function approximators were tuned simultaneously online using techniques from adaptive control to learn the value function (critic) and controller (actor) that solve an undiscounted infinite-horizon optimal control problem.[4] The method from [1] was initially limited to systems with known dynamics;

---

[4] This is the primary reason why the estimated value function weights are marked with the subscript 'c' (for critic) and the estimated policy weights are marked with the subscript 'a' (for actor).

however, extensions of this approach to uncertain systems were quickly achieved using model free [2, 3] or model-based RL methods [4]. A key limitation of early works in this area is that convergence of the weight estimates was contingent on the persistence of excitation conditions from Chap. 4, which can be seen as an analogue of the exploration paradigm often mentioned in the RL literature. The approach from [5] alleviated such a restriction using ideas from concurrent learning adaptive control [6] by introducing the idea of "simulation of experience" in which an estimated model of the system is used to generate data for learning. Similar ideas that relax the PE requirement have also been used in model free RL approaches [7]. The use of concurrent learning-based ideas in this setting often draws analogues with the idea of "experience replay" from the RL literature. Surveys of model-free RL methods stemming from the work of [1] can be found in [8, 9] whereas a collection of MBRL methods can be found in [10].

Although the online RL method from [1] was originally developed for unconstrained optimal control problems, recent works have begun to extend such ideas to those with input and state constraints. Such extensions are typically facilitated by the inclusion of additional terms in the cost function that penalize violation of such constraints. For example, works such as [7, 11–13] include a tanh-based control input penalty in the cost function, which is shown to guarantee satisfaction of hyper-rectangular actuation constraints. State constraints can be handled by including reciprocal barrier-based terms in the cost function either implicitly via coordinate transformation [14–17] or by explicitly [18–20] including terms that take large/infinite value on the boundary of the state constraint set. A challenge with including terms that take infinite values on the boundary of the constraint set is that the resulting value function may be non-differentiable at such points, which makes it challenging to approximate using smooth function approximators. A more fundamental challenge, however, is that safety guarantees in such approaches generally use the value function as a safety certificate. Since the ultimate objective of such approaches is to learn the value function, and therefore the safety certificate, learning and safety become tightly coupled—safety guarantees are contingent upon convergence of the RL algorithm, which, as discussed in this chapter, relies on conditions that cannot be verified in practice.

The approach presented in this chapter was originally introduced in [21], where a reciprocal barrier function-based controller (based on the class of Lyapunov-like barrier functions from [22]) was used as a shielding controller instead of the adaptive CBF-QP (aCBF-QP) used here. However, reciprocal barrier functions (i.e., those that take unbounded values on the boundary of the safe set) may require large control values near the boundary of the safe set and are not well-defined outside the safe set. Fortunately, the method from [21] can be easily extended to use zeroing-type barrier functions by simply replacing the reciprocal barrier function-based shielding controller from [21] with an aCBF-QP filter from Chap. 5, and is done so explicitly in this chapter as well as in [23]. Incorporating CBFs into more traditional episodic RL frameworks to endow such controllers with safety properties has also become popular recently [24, 25]. Standard references on RL include [26–28].

Lemma 8.1 is adapted from [29]. Further details on the projection operator can be found in [30, Appendix E]. Similar to Chap. 7, the system dynamics and safety constraint from Example 8.1 are taken from [31]. Further details on the universal function approximation theorem can be found in [32, 33] and [10, Chap. 2].

References

1. Vamvoudakis KG, Lewis FL (2010) Online actor-critic algorithm to solve the continuous-time infinite horizon optimal control problem. Automatica 46(5):878–888
2. Vrabie D, Pastravanu O, Abu-Khalaf M, Lewis FL (2009) Adaptive optimal control for continuous-time linear systems based on policy iteration. Automatica 45:477–484
3. Vrabie D, Lewis FL (2009) Neural network approach to continuous-time direct adaptive optimal control for partially unknown nonlinear systems. Neural Netw 22(3):237–246
4. Bhasin S, Kamalapurkar R, Johnson M, Vamvoudakis KG, Lewis FL, Dixon WE (2013) A novel actor-critic-identifier architecture for approximate optimal control of uncertain nonlinear systems. Automatica 49(1):82–92
5. Kamalapurkar R, Walters P, Dixon WE (2016) Model-based reinforcement learning for approximate optimal regulation. Automatica 64:94–104
6. Chowdhary G, Johnson E (2010) Concurrent learning for convergence in adaptive control without persistency of excitation. In: Proceedings of the IEEE conference on decision and control, pp 3674–3679
7. Modares M, Lewis FL, Naghibi-Sistani MB (2014) Integral reinforcement learning and experience replay for adaptive optimal control of partially-unknown constrained-input continuous-time systems. Automatica 50(1):193–202
8. Lewis FL, Vrabie D, Vamvoudakis KG (2012) Reinforcement learning and feedback control: Using natural decision methods to design optimal adaptive controllers. IEEE Control Syst 32(6):76–105
9. Kiumarsi B, Vamvoudakis KG, Modares H, Lewis FL (2017) Optimal and autonomous control using reinforcement learning: a survey. IEEE Trans Neural Netw Learn Syst 29(6):2042–2062
10. Kamalapurkar R, Walters P, Rosenfeld JA, Dixon WE (2018) Reinforcement learning for optimal feedback control: a lyapunov-based approach. Springer
11. Abu-Khala M, Lewis FL (2005) Nearly optimal control laws for nonlinear systems with saturating actuators using a neural network hjb approach. Automatica 41(5):779–791
12. Vamvoudakis KG, Miranda MF, Hespanha JP (2015) Asymptotically stable adaptive-optimal control algorithm with saturating actuators and relaxed persistence of excitation. IEEE Trans Neural Netw Learn Syst 27(11):2386–2398
13. Deptula P, Bell ZI, Doucette EA, Curtis JW, Dixon WE (2020) Data-based reinforcement learning approximate optimal control for an uncertain nonlinear system with control effectiveness faults. Automatica 116:1–10
14. Yang Y, Vamvoudakis KG, Modares H, He W, Yin Y, Wunsch D (2019) Safety-aware reinforcement learning framework with an actor-critic-barrier structure. In: Proceedings of the American control conference, pp 2352–2358
15. Yang Y, Vamvoudakis KG, Modares H (2020) Safe reinforcement learning for dynamical games. Int J Robust Nonlinear Control 30(9):3706–3726
16. Greene ML, Deptula P, Nivison S, Dixon WE (2020) Sparse learning-based approximate dynamic programming with barrier constraints. IEEE Control Syst Lett 4(3):743–748

17. Mahmud SMN, Hareland K, Nivison SA, Bell ZI, Kamalapurkar R (2021) A safety aware model-based reinforcement learning framework for systems with uncertainties. In: Proceedings of the American control conference, pp 1979–1984

18. Cohen MH, Belta C (2020) Approximate optimal control for safety-critical systems with control barrier functions. In: Proceedings of the IEEE conference on decision and control, pp 2062–2067

19. Marvi Z, Kiumarsi B (2021) Safe reinforcement learning: a control barrier function optimization approach. Int J Robust Nonlinear Control 31(6):1923–1940

20. Deptula P, Chen H, Licitra R, Rosenfeld JA, Dixon WE (2020) Approximate optimal motion planning to avoid unknown moving avoidance regions. IEEE Trans Robot 32(2):414–430

21. Cohen MH, Belta C (2023) Safe exploration in model-based reinforcement learning using control barrier functions. Automatica 147:110684

22. Panagou D, Stipanovic DM, Voulgaris PG (2016) Distributed coordination control for multi-robot networks using lyapunov-like barrier functions. IEEE Trans Autom Control 61(3):617–632

23. Cohen MH, Serlin Z, Leahy KJ, Belta C (2023) Temporal logic guided safe model-based reinforcement learning: a hybrid systems approach. Nonlinear Anal: Hybrid Syst 47:101295

24. Cheng R, Orosz G, Murray RM, Burdick JW (2019) End-to-end safe reinforcement learning through barrier functions for safety-critical continuous control tasks. Proc AAAI Conf Artif Intell 33:3387–3395

25. Choi J, Castaneda F, Tomlin CJ, Sreenath K (2020) Reinforcement learning for safety-critical control under model uncertainty using control lyapunov functions and control barrier functions. Robot: Sci Syst

26. Sutton RS, Barto AG (2018) Reinforcement learning: an introduction. MIT Press

27. Bertsekas DP, Tsitsiklis JN (1996) Neuro-dynamic programming. Athena Scientific

28. Bertsekas D (2019) Reinforcement learning and optimal control. Athena Scientific

29. Kamalapurkar R, Rosenfeld JA, Dixon WE (2016) Efficient model-based reinforcement learning for approximate online optimal control. Automatica 74:247–258

30. Krstić M, Kanellakopoulos I, Kokotović P (1995) Nonlinear and adaptive control design. Wiley

31. Jankovic M (2018) Robust control barrier functions for constrained stabilization of nonlinear systems. Automatica 96:359–367

32. Hornik K, Stinchcombea M, White H (1990) Universal approximation of an unknown mapping and its derivatives using multilayer feedforward networks. Neural Netw 3:551–560

33. Hornik K (1991) Approximation capabilities of multilayer feedforward networks. Neural Netw 4:251–257

Temporal Logic Guided Safe Model-Based Reinforcement Learning

9

Temporal logics are formal, expressive languages traditionally used in the computer science area of formal methods to specify the correctness of digital circuits and computer programs. Safety, as defined in previous chapters, is just a particular case of a temporal logic formula. In this chapter, we discuss adding general, temporal logic specifications to the control problem considered previously. In Sect. 9.1, we introduce Linear Temporal Logic (LTL) and automata accepting languages satisfying LTL formulas. Our approach to constructing controllers that enforce the satisfaction of an LTL formula is to break down an LTL formula into a sequence of reach-avoid subproblems. In Sect. 9.2, we formulate and solve these subproblems, and in Sect. 9.3 we present a hybrid system approach to combine the above controllers in such a way that the trajectories of the closed loop system satisfy the formula. We extend this procedure to systems for which CLFs are not known or for which the dynamics are uncertain in Sect. 9.4, where we leverage the MBRL framework from Chap. 8. We illustrate the method developed in this chapter with numerical examples in Sect. 9.5 and conclude with final remarks, references, and suggestions for further reading in Sect. 9.6.

Until this chapter, our objective was to stabilize a control system, while satisfying safety specifications and other control and state constraints. Safety can be informally stated as "nothing bad ever happens". In previous chapters, a system was considered safe if it was guaranteed to stay inside a given set i.e., the outside of this set was considered "bad". Temporal logics, such as the Linear Temporal Logic (LTL) considered in this chapter, can be used to express arbitrarily rich specifications, including liveness ("something good should eventually happen"). In this chapter, we show how to accommodate specifications given in a fragment of LTL to the control problem considered previously. As in the first part of the book, we consider nonlinear control affine systems in the form (2.10), reproduced here for

© The Author(s), under exclusive license to Springer Nature Switzerland AG 2023
M. Cohen and C. Belta, *Adaptive and Learning-Based Control of Safety-Critical Systems*,
Synthesis Lectures on Computer Science,
https://doi.org/10.1007/978-3-031-29310-8_9

convenience:

$$\dot{x} = f(x) + g(x)u.$$

9.1   Temporal Logics and Automata

An LTL formula is built from a finite set of *observations*[1] O, logical connectives, and temporal operators, such as *eventually* ($\Diamond$) and *always* ($\Box$).

Definition 9.1 (*LTL syntax*) A linear temporal logic formula φ over a finite set of observations O is recursively defined as

$$\varphi = \top \mid o \mid \varphi_1 \wedge \varphi_2 \mid \neg\varphi \mid \bigcirc\varphi \mid \varphi_1 U\varphi_2, \tag{9.1}$$

where $o \in O$ is an observation, $\varphi, \varphi_1, \varphi_2$ are LTL formulas, $\top$ is the Boolean "true," the symbols $\wedge$ and $\neg$ denote conjunction and negation, respectively, and the symbols $\bigcirc$ and U denote the temporal operators "next" and "until," respectively.

The above syntax can be used to derive other temporal operators, such as "eventually" $\Diamond$ and "always" $\Box$ as

$$\Diamond\varphi := \top U\varphi,$$
$$\Box\varphi := \neg\Diamond\neg\varphi.$$

The semantics of LTL formulas are interpreted over infinite words composed of observations[2] in O (denoted by O^ω), whose formal definition we omit here and refer the reader to the notes in Sect. 9.6 for further details.

Example 9.1 (*LTL example*) As an example of an LTL formula that expresses a control specification richer than safety, we consider a surveillance task in which a mobile robot must continuously gather data from a region of interest and bring it back to a base location while remaining safe. Ultimately, our control objectives for the robot are as follows:

- The robot should continuously gather data;
- The robot should continuously return to the base to upload the data;
- The robot should continuously recharge itself;
- The robot should avoid all dangerous areas.

---

[1] The observations in an LTL formula are typically referred to as *atomic propositions*.

[2] Traditionally, the semantics of LTL formulas are interpreted over infinite words in 2^O; however, as will become clear shortly, each $o \in O$ will correspond to mutually disjoint regions in the state space of a dynamical system. Therefore, only a single observation can evaluate to true at a given instant and there is no loss of generality in considering words only over O.

We formalize the above requirements with the LTL formula

$$\varphi = \Box(\Diamond\texttt{gather} \wedge \Diamond\texttt{base} \wedge \Diamond\texttt{recharge} \wedge \neg\texttt{danger}),$$

defined over the set of observations

$$O = \{\texttt{gather}, \texttt{base}, \texttt{recharge}, \texttt{danger}\}.$$

In plain English, the above formula reads "Always eventually gather data and always eventually return to base and always eventually recharge and always avoid dangerous regions."

The language of φ is the set of all words satisfying φ and is denoted by $\mathcal{L}(\varphi)$. For any LTL formula φ, there exists a nondeterministic Büchi Automaton with input alphabet O that accepts exactly the language of φ. Various off-the-shelf tools for converting an LTL formula into a nondeterministic Büchi Automaton are discussed in the notes later on. In this chapter, we focus on the subset of LTL formulas for which there exists a *deterministic* Büchi Automaton (DBA) accepting exactly the language of φ. In short (and informally), an LTL formula can be converted into a DBA if the formula does not contain an "eventually always" sequence of temporal operators.[3]

Definition 9.2 (*Deterministic Büchi Automaton*) A *deterministic Büchi Automaton* (DBA) is a tuple $\mathcal{A} = (Q, q_0, O, \delta_{\mathcal{A}}, Q_f)$, where Q is a finite set of states, $q_0 \in Q$ is the initial state, O is the input alphabet, $\delta_{\mathcal{A}} : Q \times O \to Q$ is a transition function, and $Q_f \subset Q$ is the set of final/accepting states.

A *run* of a DBA $\mathcal{A}$ over a word $w_o = w_o(0)w_o(1)w_o(2)\cdots \in O^\omega$ is an infinite sequence of states $w_q = w_q(0)w_q(1)w_q(2)\cdots \in Q^\omega$ such that $w_q(i+1) = \delta_{\mathcal{A}}(w_q(i), w_o(i))$. A word w_o is said to be *accepted* by $\mathcal{A}$ if the corresponding run w_q intersects with the set of accepting states Q_f infinitely often. The set of all words accepted by $\mathcal{A}$ is referred to as the *language* of $\mathcal{A}$ and is denoted by $\mathcal{L}(\mathcal{A})$.

Example 9.2 (*LTL example (continued)*) The LTL formula from the proceeding example can be translated into the DBA displayed in Fig. 9.1 with states $Q = \{q_0, q_1, q_2, q_3, q_4\}$, where the set of accepting states is $Q_f = \{q_2\}$. An input word accepted by the DBA is $w_o = (\texttt{gather}\,\texttt{base}\,\texttt{recharge})^\omega$, which produces the run $w_q = q_0(q_1q_3q_2)^\omega$.

In this chapter, each $o \in O$ in an LTL formula will correspond to a *region of interest* (ROI) $\mathcal{R}_o \subset \mathbb{R}^n$ in the state space of (2.10) satisfying the following conditions.

---

[3] Note that this is not restrictive in practice. Even though $\Diamond\Box$ resembles stability, we do not use LTL specifications for it – we enforce stability with CLFs.

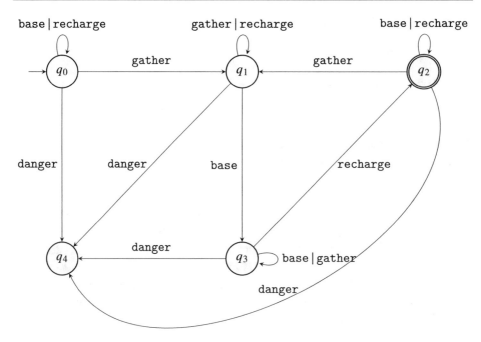

Fig. 9.1 DBA corresponding to the LTL formula from Example 9.1. If multiple transitions exist between states, only one transition, labeled by all observations that enable that transition is illustrated

Assumption 9.1 Each region of interest (ROI) $\mathcal{R}_o \subset \mathbb{R}^n$, $o \in O$ in the state space of (2.10) is closed, nonempty, contains no isolated points, and can be expressed as the zero sublevel set of a continuously differentiable function, in the sense that

$$
\begin{aligned}
\mathcal{R}_o &= \{x \in \mathbb{R}^n \mid h_o(x) \leq 0\} \\
\partial \mathcal{R}_o &= \{x \in \mathbb{R}^n \mid h_o(x) = 0\} \\
\mathrm{Int}(\mathcal{R}_o) &= \{x \in \mathbb{R}^n \mid h_o(x) < 0\}.
\end{aligned}
\tag{9.2}
$$

Moreover, all ROIs are mutually disjoint, i.e., $\mathcal{R}_o \cap \mathcal{R}_{o'} = \emptyset$ for all $(o, o') \in O \times O, o \neq o'$.

The assumption that h_o is continuously differentiable restricts the shape of each ROI (e.g., it cannot be a rectangle); however, one can typically over/under approximate such a region using an ellipse, which can be expressed using a continuously differentiable h_o. The assumption that all ROIs are disjoint implies that at most only one observation can be satisfied at a given state. To associate elements of $\mathbb{R}^n$ with the set of observations labeling the ROIs, we introduce a labeling function $\ell : \mathbb{R}^n \to O \cup \{\mathtt{null}\}$ such that

$$
\ell(x) = \begin{cases} o, & \text{if } x \in \mathcal{R}_o, \\ \mathtt{null}, & \text{otherwise.} \end{cases}
\tag{9.3}
$$

This labeling function induces an inverse map $\ell^{-1}(o) = \{x \in \mathbb{R}^n \mid \ell(x) = o\}$. The element null acts as a null observation in the sense that $\ell^{-1}(\text{null})$ corresponds to the space between the disjoint ROI, which will be useful in defining the semantics related to continuous-time trajectories. We now formalize what it means for a trajectory of (2.10) to satisfy an LTL formula φ.

Definition 9.3 (*Semantics of a continuous trajectory*) Let $x : \mathbb{R}_{\geq 0} \to \mathbb{R}^n$ be a continuous curve. The word generated by $x(\cdot)$ is a sequence $w_x = w_x(0)w_x(1)w_x(2)\ldots$ recursively defined as

1. $w_x(0) = \ell(x(0))$;
2. $w_x(i) = \lim_{t \to \tau_i^+} \ell(x(t))$ for all $i \geq 1$ such that $\tau_i < \infty$, where τ_i is defined as $\tau_0 = 0$ and $\tau_i := \inf\{t \mid t > \tau_{i-1}, \ell(x(t)) \neq w_x(i-1)\}$ for all $i \geq 1$;
3. $w_x(i) = w_x(i-1)$ for all i such that $\tau_i = \infty$.

The word w_x is said to be well-defined if $\tau_i \to \infty$ as $i \to \infty$. The curve $x(\cdot)$ is said to satisfy an LTL formula φ if and only if the word it generates is well-defined and $\text{proj}_O(w_x) \subset \mathcal{L}(\varphi)$, where $\text{proj}_O(w_x)$ removes all instances of null from w_x.

The above definition implies that the word generated by a continuous curve is exactly the sequence of observations corresponding to the regions intersected by the curve over time, where each observation is generated at the instant the curve switches from one region to another.

9.2   Simultaneous Stabilization and Safety

Our approach to constructing controllers that enforce the satisfaction of an LTL formula by the closed-loop trajectory of (2.10) is to break down an LTL formula into a sequence of reach-avoid sub-problems dictated by the structure of the formula's corresponding DBA. Central to this approach is the ability to construct continuous controllers that allow for stabilizing (2.10) to various ROI while avoiding those that should not be visited. In previous chapters, we have constructed controllers for (2.10) enforcing stability and safety using CLFs and CBFs, respectively, by filtering a nominal stabilizing controller k_0 through the CBF-QP (3.12)

$$k(x) = \operatorname*{argmin}_{u \in \mathbb{R}^m} \tfrac{1}{2}\|u - k_0(x)\|^2$$
$$\text{subject to} \quad L_f h(x) + L_g h(x)u \geq -\alpha(h(x)),$$

where $h : \mathbb{R}^n \to \mathbb{R}$ is a CBF for (2.10) on a set $C \subset \mathbb{R}^n$ as in (3.3) and $\alpha \in \mathcal{K}_\infty^e$. Recall that, provided all the functions involved in (3.12) are Lipschitz continuous, the solution to (3.12) is as well, and can be expressed in closed-form as

$$k(x) = \begin{cases} k_0(x) & \text{if } \psi(x) \geq 0 \\ k_0(x) - \frac{\psi(x)}{\|L_g h(x)^\top\|^2} L_g h(x)^\top & \text{if } \psi(x) < 0, \end{cases} \tag{9.4}$$

where $\psi(x) := L_f h(x) + L_g h(x) k_0(x) + \alpha(h(x))$. Empirically, the above controller performs well at enforcing both (asymptotic) stability and safety; however, as stability is treated as a "soft" constraint (encoded through the cost function) in the CBF-QP, there is no formal guarantee, in general, that such a controller enforces asymptotic stability of the origin. The question we provide an answer to in this section is as follows: When the nominal policy k_0 enforces asymptotic stability of the closed-loop system using a Lyapunov function V, when is V also a Lyapunov function for the closed-loop system under the CBF-QP controller (3.12) using k_0 as the nominal policy? We provide a (conservative) answer to this question using the notion of a *CBF-stabilizable set*.

Definition 9.4 (*CBF stabilizable set*) Let h be a CBF and V a CLF for (2.10) with an associated locally Lipschitz controller $k_0 : \mathbb{R}^n \to \mathbb{R}^m$ satisfying

$$L_f V(x) + L_g V(x) k_0(x) \leq -\gamma(V(x)), \quad \forall x \in \mathbb{R}^n, \tag{9.5}$$

where $\gamma \in \mathcal{K}$. Then, the set

$$\mathcal{V}_l := \{x \in \mathbb{R}^n \mid V(x) \leq l\} \tag{9.6}$$

is said to be *CBF-stabilizable* if either one of the following conditions hold:

1. For all $x \in C \cap \mathcal{V}_l$, we have

$$L_g V(x) L_g h(x)^\top \leq 0. \tag{9.7}$$

2. Condition (9.7) only holds on some subset $\mathcal{S} \subset C \cap \mathcal{V}_l$, and everywhere on the complement of such a subset, we have

$$\underbrace{L_f h(x) + L_g h(x) k_0(x) + \alpha(h(x))}_{\psi(x)} \geq 0, \quad \forall x \in (C \cap \mathcal{V}_l) \setminus \mathcal{S}. \tag{9.8}$$

Given a CBF-stabilizable set, the region of points where (9.7) holds represents the set of all points where the CBF does not interfere with the stabilization objective. The second condition from Definition 9.4 allows for the existence of points where the CBF could act to prevent stabilization, but where the nominal controller k_0 satisfies the CBF conditions without any intervention and so such interference is not necessary. To study stability properties of the closed-loop system under the CBF-QP controller, we must first ensure that the origin is indeed an equilibrium point of the closed-loop system.

Lemma 9.1 *Consider system (2.10) with $f(0) = 0$ and a locally Lipschitz nominal controller $k_0 : \mathbb{R}^n \to \mathbb{R}^m$ satisfying $k_0(0) = 0$. Let $h : \mathbb{R}^n \to \mathbb{R}$ be a CBF for (2.10) on a* *set* $C \subset \mathbb{R}^n$ *as in (3.3) with* $0 \in Int(C)$. *Then, the origin is an equilibrium point for the closed-loop system under the CBF-QP controller (3.12).*

Proof Under the assumptions of the lemma, we have that

$$\psi(0) = \underbrace{L_f h(0)}_{=0} + \underbrace{L_g h(0) k_0(0)}_{=0} + \underbrace{\alpha(h(0))}_{>0} > 0.$$

It then follows from (9.4) that $k(0) = k_0(0)$, which implies that $f(0) + g(0)k(0) = 0$, so that the origin is an equilibrium point for the closed-loop system, as desired. ☐

The following lemma will be helpful in proving the next result.

Lemma 9.2 *Let* $f : \mathbb{R}^n \to \mathbb{R}^n$ *be a locally Lipschitz vector field and let* $\Omega \subset \mathbb{R}^n$ *be a compact set. If the solution of the initial value problem*

$$\dot{x}(t) = f(x(t)), \quad \forall t \in I(x_0),$$
$$x(0) = x_0,$$

satisfies $x(t) \in \Omega$ *for all* $t \in I(x_0)$, *where* $I(x_0) = [0, \tau_{\max})$ *is the solution's maximal interval of existence from an initial condition of* x_0, *then* $\tau_{\max} = \infty$.

We now show that the existence of a CBF-stabilizable set is sufficient to ensure asymptotic stability of the origin under the CBF-QP controller.

Theorem 9.1 *Suppose the conditions of Lemma 9.1 hold and let* $\mathcal{V}_l \subset \mathbb{R}^n$ *be a CBF-stabilizable set. Then,* $\mathcal{V}_l \cap C$ *is forward invariant for the closed-loop system, closed-loop trajectories exist for all time, and the origin is asymptotically stable for the closed-loop system.*

Proof It follows from Lemma 9.1 that the origin is an equilibrium point for the closed-loop system. Now take the CLF V from Definition 9.4 as a Lyapunov function candidate for the closed-loop system. The Lie derivative of V along the closed-loop vector field is

$$\dot{V}(x) = L_f V(x) + L_g V(x) k(x)$$

$$= \begin{cases} L_f V(x) + L_g V(x) k_0(x), & \text{if } \psi(x) \geq 0, \quad (9.9) \\ L_f V(x) + L_g V(x) k_0(x) - \frac{\psi(x)}{\|L_g h(x)^\top\|^2} L_g V(x) L_g h(x)^\top, & \text{if } \psi(x) < 0. \end{cases}$$

Our objective is now to show that $\dot{V}(x) < 0$ for all $x \in (\mathcal{V}_l \cap C) \setminus \{0\}$. If $\mathcal{V}_l$ is CBF-stabilizable and (9.7) holds for all $x \in \mathcal{V}_l \cap C$, then (9.9) in conjunction with (9.5) implies that $\dot{V}(x) \leq -\gamma(V(x))$ for all $x \in \mathcal{V}_l \cap C$. On the other hand, if (9.7) only holds on some subset $S \subset \mathcal{V}_l \cap C$ but for all points in $(\mathcal{V}_l \cap C) \setminus S$ we have $\psi(x) \geq 0$, then $\dot{V}(x) \leq -\gamma(V(x))$ still holds for all $x \in \mathcal{V}_l \cap C$ by (9.9). Thus, in either case, if $\mathcal{V}_l$ is CBF-stabilizable, we have

$$\dot{V}(x) \leq -\gamma(V(x)), \quad \forall x \in \mathcal{V}_l \cap C. \tag{9.10}$$

We now show that $\mathcal{V}_l \cap C$ is forward invariant for the closed-loop system and that system trajectories are defined for all time. As $\dot{V}(x) \leq 0$ for all $x \in \mathcal{V}_l \cap C$, then $\dot{V}(x) \leq 0$ for any sublevel set of V contained in $\mathcal{V}_l \cap C$. This implies that a given trajectory $t \mapsto x(t)$ cannot exit $\mathcal{V}_l \cap C$ via $\partial \mathcal{V}_l$ or via ∂C since h is a CBF and the controller in (9.4) ensures $\dot{h}(x) \geq 0$ for all $x \in \partial C$. As the closed-loop vector field is locally Lipschitz, given an initial condition $x_0 \in \mathcal{V}_l \cap C$ there exists a maximal interval of existence $I(x_0) = [0, \tau_{\max})$ and a continuously differentiable trajectory $x : I(x_0) \to \mathbb{R}^n$ solving the initial value problem

$$\dot{x}(t) = f(x(t)) + g(x(t))k(x(t)), \quad \forall t \in I(x_0),$$
$$x(0) = x_0.$$

Based on the preceding argument, the solution of the initial value problem satisfies $x(t) \in \mathcal{V}_l \cap C$ for all $t \in I(x_0)$, which implies $\mathcal{V}_l \cap C$ is forward invariant for the closed-loop system. To show that $\tau_{\max} = \infty$ in the above initial value problem, we note that since C is closed and $\mathcal{V}_l$ is compact, $\mathcal{V}_l \cap C$ is a compact set, and it follows from Lemma 9.2 that $\tau_{\max} = \infty$. As the solution is defined for all times and, along such a solution, we have $\dot{V}(x(t)) \leq -\gamma(V(x(t))$, V is a Lyapunov function for the closed-loop system and the origin is asymptotically stable. $\qquad \square$

Provided $0 \in \text{Int}(C)$, h is a CBF, and V is a CLF, it is always possible to find a CBF stabilizable set simply by taking l from (9.6) as an arbitrarily small positive constant. Clearly, such an approach is highly conservative and finding large CBF-stabilizable sets is a challenging problem. Despite the conservatism of these theoretical results, we demonstrate empirically later on that simultaneous stabilization and safety can be achieved even in rather complex environments. In what follows, we are not necessarily interested in stabilizing (2.10) to a particular point asymptotically; rather, we are interested in ensuring that trajectories of the closed-loop system reach a desired region in finite time. Fortunately, such a problem can be addressed by rendering a point on the interior of such a set *uniformly*[4] asymptotically stable.

Proposition 9.1 *Let $\mathcal{T} \subset \mathbb{R}^n$ be a non-empty set containing no isolated points and let $0 \in \text{Int}(\mathcal{T})$. Provided the conditions of Theorem 9.1 holds, then there exists a finite $T \in \mathbb{R}_{\geq 0}$ such that $x(T) \in \mathcal{T}$.*

---

[4] Asymptotic stability is equivalent to uniform asymptotic stability for time-invariant systems.

Proof Since $0 \in \text{Int}(C)$, $0 \in \text{Int}(\mathcal{V}_l)$, and $0 \in \text{Int}(\mathcal{T})$ then $0 \in \text{Int}(C) \cap \text{Int}(\mathcal{V}_l) \cap \text{Int}(\mathcal{T})$. As the intersection of a finite number of set interiors is equal to the interior of the set intersection we have $0 \in \text{Int}(C \cap \mathcal{V}_l \cap \mathcal{T})$. As the origin is contained in the interior of $C \cap \mathcal{V}_l \cap \mathcal{T}$, then there must exist a $\delta \in \mathbb{R}_{>0}$ such that $\mathcal{B}_\delta(0) \subset C \cap \mathcal{V}_l \cap \mathcal{T}$. Now note that the origin of the closed-loop system is asymptotically stable by Theorem 9.1. Since the closed-loop system is time-invariant, this implies that the origin is uniformly asymptotically stable, which implies that for each $\delta' \in \mathbb{R}_{>0}$ there exists a time $T = T(\delta')$ such that $x(t) \in \mathcal{B}_{\delta'}(0)$ for all $t \geq T$. Taking $\delta' < \delta$, we have $\mathcal{B}_{\delta'}(0) \subset \mathcal{B}_\delta(0)$, which implies that $x(t)$ enters $\mathcal{B}_\delta(0)$ by time $T < \infty$, which also implies $x(t)$ enters $\mathcal{T}$ by time $T < \infty$, as desired.   □

Remark 9.1 Although the results of this section are tailored to the special case in which the origin is the desired equilibrium point to be asymptotically stabilized, the same ideas are applicable to an arbitrary equilibrium point via a coordinate transformation on the state and, in certain cases, the control input. An explicit example of this is provided in Chap. 9.4.

9.3   A Hybrid Systems Approach to LTL Control Synthesis

In this section, we illustrate how the CBF controller from the previous section can be used in a hybrid system framework to synthesize a control policy capable of satisfying an LTL formula. To this end, consider an LTL formula φ defined over a finite set of observations O corresponding to a collection of ROIs satisfying Assumption 9.1. Given a CBF stabilizable set and a CBF-QP-based controller capable of steering (2.10) to a target set $\mathcal{T}$ while remaining within a safe set C, our objective is to coordinate the sequence of sets that are reached/avoided by (2.10) in order to generate a word that satisfies the specification. To formally specify the switching logic used to select among a family of feedback controllers for (2.10), we consider augmenting the continuous dynamics with the discrete dynamics of the DBA corresponding to φ to form a hybrid control system. To enforce satisfaction of the specification over the resulting product hybrid system, we leverage the notion of the *distance to acceptance* function (DTA). The DTA function is, in essence, a Lyapunov-like function for a given DBA $\mathcal{A} = (Q, q_0, O, \delta_{\mathcal{A}}, Q_f)$ corresponding to an LTL formula φ in that it is non-negative for all $q \in Q$, positive for all $q \in Q \setminus Q_f$, and zero for all $q \in Q_f$. Rather than enforcing convergence of trajectories to an equilibrium, however, the DTA will be used to enforce convergence to the set of accepting states infinitely often.

To formalize the preceding discussion, let $\mathcal{A} = (Q, q_0, O, \delta_{\mathcal{A}}, Q_f)$ be a DBA corresponding to φ and let $\mathcal{P}(q_i, q_j)$ denote the set of all paths in $\mathcal{A}$ from $q_i \in Q$ to $q_j \in Q$. More formally

$$\mathcal{P}(q_i, q_j) := \{q_1 \ldots q_n \mid q_1 = q_i, \, q_n = q_j, \, \forall k \in [1, n-1], \\ \exists o \in O \text{ s.t. } q_{k+1} = \delta_{\mathcal{A}}(q_k, o)\}. \tag{9.11}$$

The distance between any two states in $\mathcal{A}$ is then defined as

$$d(q, q') := \begin{cases} \min_{\mathbf{q} \in \mathcal{P}(q,q')} \Upsilon(\mathbf{q}), & \text{if } \mathcal{P}(q, q') \neq \emptyset, \\ \infty, & \text{if } \mathcal{P}(q, q') = \emptyset, \end{cases} \qquad (9.12)$$

where $\mathbf{q} \in \mathcal{P}(q, q')$ denotes a path from q to q' and $\Upsilon(\mathbf{q})$ denotes the number of states in $\mathbf{q}$. The DTA can then be computed as

$$V_d(q) := \min_{q' \in Q_f} d(q, q'), \qquad (9.13)$$

and represents the minimum number of transitions needed to reach an accepting state from any given $q \in Q$. These properties imply that enforcing satisfaction of a given LTL formula φ is equivalent to ensuring runs of $\mathcal{A}$ reach a set of states where $V_d(q) = 0$ infinitely often. The following results outline some useful properties of the DTA.

Lemma 9.3 *The distance to acceptance (DTA) function from (9.13) satisfies the following properties:*

(i) *An accepting run w_q of $\mathcal{A}$ cannot contain a state $q \in Q$ such that $V_d(q) = \infty$.*
(ii) *For each $q \in Q$, if $V(q) > 0$ and $V(q) \neq \infty$, then there exists a state q' and an observation $o \in O$ such that $q' = \delta_{\mathcal{A}}(q, o)$ and $V_d(q') < V_d(q)$.*

Proof (i) Recall that a run $w_q = w_q(0)w_q(1)w_q(2)\cdots \in Q^\omega$ is accepting if it intersects with Q_f infinitely many times. For the sake of contradiction, suppose there exists a $j \in \mathbb{N}$ such that $V_d(w_q(j)) = \infty$. Based on (9.12) and (9.13) this implies that $\mathcal{P}(w_q(j), q') = \emptyset$ for all $q' \in Q_f$. That is, there exists no path in $\mathcal{A}$ starting from $w_q(j)$ and ending in an accepting state. Thus, there exists no $j' > j$ such that $w_q(j') \in Q_f$, which contradicts the initial assumption that w_q is an accepting run of $\mathcal{A}$. Hence, there cannot exist an accepting run containing a state $q \in Q$ such that $V_d(q) = \infty$, and (i) follows.

(ii) Since $q \in Q$ satisfies $V(q) > 0$, then $q \notin Q_f$. Moreover, since $V_d(q) \neq \infty$, then (9.12) implies there exists a finite shortest path $q_1 q_2 \dots q_n$ such that $q_1 = q$ and $q_n \in Q_f$. It follows from Bellman's Principle of Optimality that the finite path $q_2 \dots q_n$ is the shortest path starting at q_2 to reach a state in Q_f, hence

$$V_d(q) = d(q, q_2) + V_d(q_2).$$

Since $d(q, q_2) > 0$, the preceding equality only holds if $V_d(q_2) < V_d(q)$. Since $\mathcal{P}(q, q_2) \neq \emptyset$, with $\mathcal{P}$ as in (9.11), then there must exist an observation $o \in O$ such that $q_2 = \delta_{\mathcal{A}}(q, o)$, and (ii) follows from taking $q' = q_2$. $\square$

The preceding lemma implies that for any $q \in Q \backslash Q_f$ such that $V_d(q) \neq \infty$, there always exists a $q' \in Q$ such that $V_d(q') < V_d(q)$. This observation allows us to define the set $O_q := \{o \in O \mid \exists q' \in Q \text{ s.t. } q' = \delta_{\mathcal{A}}(q, o)\}$, which is the set of all admissible observations for state

q, and

$$\bar{O}_q := \operatorname{argmin}_{o \in O_q} V_d(\delta_{\mathcal{A}}(q, o)). \tag{9.14}$$

For each $q \in Q$, the set $\bar{O}_q$ in (9.14) represents the set of all observations that force a transition to the reachable state with the minimum DTA. For $q \in Q \backslash Q_f$ this guarantees that $V_d(q') < V(q)$, where $q' = \delta_{\mathcal{A}}(q, o)$ and $o \in \bar{O}_q$, whereas for $q \in Q_f$ inputting observation $o \in \bar{O}_q$ forces a transition to the reachable state that incurs the smallest increase in DTA. We now leverage the properties of the DTA to construct a product hybrid system that captures the continuous behavior (2.10) and the discrete behavior of $\mathcal{A}$. To formalize this idea, we first introduce the notion of a hybrid system used in this chapter:

Definition 9.5 (*Hybrid system*) A hybrid system $\mathcal{H}$ is a tuple

$$\mathcal{H} = (Q, X, \text{Init}, \text{Dom}, \mathcal{F}, E, \text{Guard}),$$

where

- Q is a finite set of discrete modes;
- $X \subset \mathbb{R}^n$ is a set of continuous states;
- Init $\subset Q \times X$ is a set of initial states;
- Dom : $Q \rightrightarrows X$ is a set-valued map that associates to each discrete mode a set $\text{Dom}(q) \subset X$ that describes the set of admissible continuous states while in mode q;
- $\mathcal{F} : Q \times X \to \mathbb{R}^n$ is a mapping that associates to each discrete mode $q \in Q$ a vector field $\mathcal{F}_q : X \to \mathbb{R}^n$ that characterizes the evolution of the continuous states;
- $E \subset Q \times Q$ is a set of edges that describe the admissible transitions between discrete modes.

Next, we define the semantics of a hybrid system:

Definition 9.6 (*Semantics of a hybrid system*) An execution of $\mathcal{H}$ is interpreted over a hybrid time set, which is a finite or infinite sequence of intervals of the form $\tau = \{I_i\}_{i=0}^N$ such that: $I_i = [\tau_i, \tau_i']$ for all $i < N$; if $N < \infty$, then I_N may be right open or closed; and $\tau_i \le \tau_i' = \tau_{i+1}$ for all i, where each τ_i represents the times at which discrete transitions take place. Given a hybrid time set τ, we denote by $\langle \tau \rangle$ the set $\{0, 1, \ldots, N\}$ if $N < \infty$ and $\{0, 1, \ldots\}$ if $N = \infty$. Given an initial time $t_0 \in \mathbb{R}_{\ge 0}$, an execution of $\mathcal{H}$ is a collection $\xi = (\tau, q, x)$, where τ is a hybrid time set, $q : \langle \tau \rangle \to Q$ is a mapping from $\langle \tau \rangle$ to the set of discrete modes, and $x = \{x^i : i \in \langle \tau \rangle\}$ is a collection of differentiable maps $x^i : I_i \to X$, such that: $(q(0), x^0(t_0)) \in \text{Init}$; for all $t \in [\tau_i, \tau_i')$ we have $x^i(t) \in \text{Dom}(q(i))$ and $\dot{x}^i(t) = \mathcal{F}_{q(i)}(x^i(t))$; for all $i \in \langle \tau \rangle \backslash \{N\}$ we have $(q(i), q(i+1)) \in E$ and $x^i(\tau_i') \in \text{Guard}(q(i), q(i+1))$.

To study the continuous system and the discrete DBA corresponding to an LTL formula in a unified framework, let $\mathcal{H} = (Q, X, \text{Init}, \text{Dom}, \mathcal{F}, E, \text{Guard})$ be a hybrid system as in Definition 9.5, where

- Q is the set of discrete modes inherited from $\mathcal{A}$;
- $X = \mathbb{R}^n$ is the continuous state space;
- $\text{Init} = \{q_0\} \times X$ is the set of initial states, where q_0 is inherited from $\mathcal{A}$;
- $\text{Dom}(q) = X \setminus \cup_{o \in O_q} \ell^{-1}(o)$ is the set of all states in X that lie outside regions of interest corresponding to observations in O_q;
- $\mathcal{F}(q, x) = f(x) + g(x)k(q, x)$, where f and g are vector fields from (2.10) and $k : Q \times X \to \mathbb{R}^m$ is a hybrid feedback law to be specified;
- $E = \{(q, q') \in Q \times Q \mid \exists o \in O_q \text{ s.t. } q' = \delta_{\mathcal{A}}(q, o)\}$ is a set of edges such that a transition from mode q to mode q' can be taken if there exists an observation from O_q enabling such a transition in $\mathcal{A}$;
- $\text{Guard}(q, q') = \{x \in \ell^{-1}(o) \mid q' = \delta_{\mathcal{A}}(q, o)\}$ is the ROI corresponding observation o such that $q' = \delta_{\mathcal{A}}(q, o)$.

The interpretation of the above definition is that for each $q \in Q$ the continuous state evolves within the domain $\text{Dom}(q)$ until $\text{Guard}(q, q')$ is reached, at which point a discrete transition governed by the dynamics of $\mathcal{A}$ takes place. Unfortunately, the structure of $\mathcal{H}$ does not take into account unsafe regions that correspond to observations that should *not* appear in the accepting word w_o. To address this limitation we associate with each $q \in Q$ a safe set $C_q \subset X$ that the continuous state must remain in while in mode q. To formalize this notion, consider the set $O_q^- := O_q \backslash \bar{O}_q$, which is the set of all observations that do not minimize the DTA when transitioning out of the current mode. For each $o \in O_q^-$ define the set $C_{q,o} := X \backslash \text{Int}(\mathcal{R}_o) = \{x \in X \mid h_o(x) \geq 0\}$, which is used to construct the overall safe set for mode q as $C_q = \cap_{o \in O_q^-} C_{q,o}$ with the convention that points on the boundary of such ROI are considered safe.[5] To ensure the safe set associated with each $q \in Q$ can be rendered forward invariant we make the following assumption.

Assumption 9.2 For each $q \in Q$ the set $O_q^- = \{o_q^-\}$ is a singleton and $h_{o_q^-}$ is a CBF for (2.10) over C_q.

The above assumption is motivated by the fact that results regarding CBF-stabilizable sets are developed for only a single CBF. This assumption may appear very restrictive at first glance - after all, we would likely need to consider multiple safe sets given a complex LTL specification. However, multiple CBFs can be combined smoothly using an under approximation of the min operator to obtain a single CBF that captures multiple safe sets. To this end, note that multiple barrier functions can be combined in a *nonsmooth* fashion as

---

[5] If this is undesirable in practice one can always take $C_{q,o} := \{x \in X \mid h_o(x) \geq \epsilon\}$ for some $\epsilon \in \mathbb{R}_{>0}$.

$$h_1 \wedge h_2 = \min\{h_1, h_2\},$$

where h_1 and h_2 are CBFs. To avoid the use of nonsmooth analysis, given a collection of CBFs $h_i, i \in \{1, \dots, N\}$, the min operator can be under-approximated as

$$- \ln \left(\sum_{i=1}^{N} \exp(-h_i(x)) \right) \leq \min_{i \in \{1, \dots, N\}} h_i(x), \qquad (9.15)$$

which provides a conservative sufficient condition for the satisfaction of multiple CBF constraints. Of course, it may be challenging to verify that the above smooth combination of CBFs is a CBF in its own right, and for the remainder of this chapter we leverage Assumption 9.2 to ensure that such a smooth combination indeed produces a CBF.

To specify a control policy for the closed-loop vector field associated with each $q \in Q$, let $\{\Phi_q\}_{q \in Q}$ be a collection of diffeomorphisms such that $\Phi_q(x) := x - x_d(q)$, where $x_d :$ $Q \rightarrow \text{Int}(\mathcal{R}_o)$ with $o \in \bar{O}_q$ is a mapping that associates to each $q \in Q$ a point on the interior of the region that should be visited by the continuous trajectory of $\mathcal{H}$ to enable a transition to a mode with the minimum DTA.[6] To ensure that the continuous trajectory of $\mathcal{H}$ is regulated to $\mathcal{R}_o$ for $o \in \bar{O}_q$, we associate to each $q \in Q$ a CLF V_q satisfying

$$\gamma_{1,q}(\|\Phi_q(x)\|) \leq V_q(\Phi_q(x)) \leq \gamma_{2,q}(\|\Phi_q(x)\|),$$

for some $\gamma_1, \gamma_2 \in \mathcal{K}_\infty$, a locally Lipschitz controller $k_{0,q}$ satisfying

$$L_f V_q(\Phi_q(x)) + L_g V_q(\Phi_q(x)) k_{0,q}(\Phi_q(x)) \leq -\gamma_{3,q}(V_q(\Phi_q(x))), \quad \forall x \in \mathbb{R}^n, \quad (9.16)$$

and a CBF-QP as in (3.12) that filters the nominal policy $k_{0,q}$ to ensure the system stays in the safe set associated with mode q. Now let $\{V_q\}_{q \in Q}$ be the collection of CLFs associated with each mode and for each $q \in Q$, $l_q \in \mathbb{R}_{>0}$, consider the set $\mathcal{V}_{l,q} := \{x \in \mathbb{R}^n \mid V_q(\Phi(x)) \leq l_q\}$. We now make the following assumption to ensure that the definition of $\mathcal{H}$ is well-posed in the sense that there exists a suitable continuous feedback law that drives the continuous trajectory of $\mathcal{H}$ from one ROI to another while remaining safe.

Assumption 9.3 Given an execution of $\mathcal{H}, \xi = (\tau, q, x)$, the set $\mathcal{V}_{l,q(i)}$ is CBF-stabilizable for all $i \in \mathbb{Z}_{\geq 0}$ and $x^i(\tau_i) \in \mathcal{V}_{l,q(i)} \cap C_{q(i)}$ for all $i \in \mathbb{Z}_{\geq 0}$.

The above assumption ensures that there exists a CBF-stabilizable set for each mode which the hybrid execution traverses and that, upon transitioning to each mode, the trajectory is contained in the intersection of the CBF-stabilizable set and the safe set. The controller ultimately applied to the system is

---

[6] In the case when $\bar{O}_q$ is not a singleton the particular $o \in \bar{O}_q$ used to define $q \mapsto x_d(q)$ can be chosen arbitrarily from $\bar{O}_q$.

$$k_q(x) = \operatorname{argmin}_{u \in \mathbb{R}^m} \ \tfrac{1}{2} \|u - k_{0,q}(x)\|^2$$
$$\text{subject to} \quad L_f h_{o_q^-}(x) + L_g h_{o_q^-}(x) u \geq -\alpha(h_{o_q^-}(x)), \tag{9.17}$$

where $k_{0,q}$ is the nominal hybrid CLF policy from (9.16) that drives the trajectory to the desired ROI and $h_{o_q^-}$ is the CBF from Assumption 9.2 that represents the safe set associated with mode q. The following proposition shows that if there exists a suitable sequence of CBF-stabilizable sets associated with the sequence of stabilization/safety problems dictated by the structure of $\mathcal{H}$, then the word generated by the execution of $\mathcal{H}$ under the hybrid CBF-based policy (9.17) satisfies φ.

Proposition 9.2 *Consider system (2.10), an LTL formula φ defined over a finite set of observations O, a hybrid product system $\mathcal{H}$ as in Definition 9.5, and assume there exists an accepting run of $\mathcal{A}$. Provided Assumptions 9.1–9.3 hold, then the word produced by the execution of $\mathcal{H}$ under control (9.17) is accepted by $\mathcal{A}$.*

Proof We start by noting that any accepting run of $\mathcal{A}$ cannot contain a mode $q \in Q$ for which $V_d(q) = \infty$ by Lemma 9.3. Since $\mathcal{A}$ is deterministic and there is only one possible initial state, then, by assumption, $V_d(q(0)) < \infty$. Hence, by Lemma 9.3, there exists some $o \in O$ such that $V_d(\delta_{\mathcal{A}}(q(0), o)) < V_d(q(0))$. Provided $x^0(\tau_0) \in \mathcal{V}_{l,q(0)} \cap C_{q(0)}$ and $\mathcal{V}_{l,q(0)}$ is CBF-stabilizable, then Theorem 9.1 and Proposition 9.1 imply $x^0(t)$ reaches $\text{Guard}(q(0), q')$ for some q' such that $V_d(q') < V_d(q(0))$ in finite time and remains in $C_{q(0)}$ for all times beforehand. Noting that Q is a finite set and repeatedly applying the same argument under the hypothesis that $x^i(\tau_i) \in \mathcal{V}_{l,q(i)} \cap C_q(i)$ and $\mathcal{V}_{l,q(i)}$ is CBF-stabilizable for all $i \in \mathbb{Z}_{\geq 0}$ implies that there exists a finite sequence of modes $q(0), q(1), \ldots, q(\bar{N})$ such that $V_d(q(0)) > V_d(q(1)) > \cdots > V_d(q(\bar{N})) = 0$. The assumption that $\mathcal{V}_{l,q(i)}$ is CBF-stabilization for each i implies that the previous transitions occur without the continuous trajectory generating any observations that should not appear in an accepting word. Since it is assumed there exists an accepting run of $\mathcal{A}$ and such a run cannot contain any states with an infinite DTA, there must exist some $o \in O$ such that $V_d(\delta_{\mathcal{A}}(q(\bar{N})), o) < \infty$. Under the hypothesis that $\mathcal{V}_{l,q(\bar{N})}$ is CBF-stabilizable and that $x^{\bar{N}}(\tau_{\bar{N}}) \in \mathcal{V}_{l,q(\bar{N})} \cap C_{q(\bar{N})}$, then Theorem 9.1 and Proposition 9.1 again imply $x^{\bar{N}}(t') \in \text{Guard}(q(\bar{N}), q(\bar{N}+1))$ for some finite t'. Repeatedly applying the same argument using $q(\bar{N}+1)$ as the initial mode implies that modes with $V_d(q) = 0$ are visiting infinitely often while always remaining in the corresponding safe set, which, by Definition 9.3, implies satisfaction of φ. $\square$

9.4 Temporal Logic Guided Reinforcement Learning

In the previous section, we illustrated how controllers based on CBFs and CLFs could be coordinated to solve a sequence of reach-avoid problems towards ultimately satisfying an LTL specification. A significant limitation of this approach is that it requires constructing a potentially large number of CBFs and CLFs to accomplish the LTL objective. In this section, we relax the requirement of knowledge of CLFs for the stabilization tasks by leveraging the model-based reinforcement learning (MBRL) framework from Chap. 8 to solve a sequence of optimal control problems (whose associated values functions are CLFs) to obtain stabilizing control policies. Shifting to this MBRL setting also allows for considering systems with uncertain dynamics using the adaptive CBFs (aCBFs) from Chap. 5.

To extend the framework of Chap. 8 to this setting, we first demonstrate how such a framework can be used to steer the trajectory to a desired target set $\mathcal{T} \subset \mathbb{R}^n$ while remaining in a safe set $C \subset \mathbb{R}^n$. To this end, we consider the uncertain system from (4.1)

$$\dot{x} = f(x) + F(x)\theta + g(x)u,$$

restated above for convenience. We make the following assumption on the control directions to ensure that the above system can be stabilized to arbitrary points.

Assumption 9.4 The matrix of control directions g is full row rank for all $x \in \mathbb{R}^n$, which guarantees existence of the Moore-Penrose pseudo-inverse $g^{\dagger}(x) := g(x)^{\top}(g(x)g(x)^{\top})^{-1}$ for all $x \in \mathbb{R}^n$.

Under Assumption 9.4, given a desired state $x_d \in \mathcal{T}$, where $\mathcal{T} \subset \mathbb{R}^n$ is a target set we wish to reach, selecting $u_d := -g^{\dagger}(x_d)(f(x_d) + F(x_d)\theta)$ ensures x_d is an equilibrium point for $\dot{x} = f(x) + F(x)\theta + g(x)u_d$. Given the feed-forward input u_d, we consider decomposing the control for (4.1) as $u = u_d + \mu$, where $\mu : \mathbb{R}^n \to \mathbb{R}^m$ is a feedback law to be determined that regulates (4.1) to $\mathcal{T}$. To facilitate this approach, define the regulation error $\eta := x - x_d$, let $\Phi : \mathbb{R}^n \to \mathbb{R}^n$ be a diffeomorphism such that $\eta = \Phi(x)$ and $x = \Phi^{-1}(\eta)$, and consider the auxiliary dynamical system

$$\dot{\eta} = \mathfrak{f}(\eta) + \mathfrak{F}(\eta)\theta + \mathfrak{g}(\eta)\mu, \tag{9.18}$$

where

$$\mathfrak{f}(\eta) := f(\Phi^{-1}(\eta)) - g(\Phi^{-1}(\eta))g(x_d)f(x_d)$$
$$\mathfrak{F}(\eta) := F(\Phi^{-1}(\eta)) - g(\Phi^{-1}(\eta))g(x_d)F(x_d)$$
$$\mathfrak{g}(\eta) := g(\Phi^{-1}(\eta)).$$

Note that $\dot{\eta} = \dot{x} - \dot{x}_d = \dot{x}$ so that trajectories $t \mapsto x(t)$ of (4.1) can be uniquely recovered from trajectories $t \mapsto \eta(t)$ of (9.18) as $x(t) = \Phi^{-1}(\eta(t))$. Since, under Assumption 9.4, system (4.1) can always be put into the form of (9.18) to achieve stabilization to a point other than the origin, in what follows our development will focus on (4.1) for ease of exposition. As a general approach to obtaining a stabilizing controller and stability certificate, we follow the same procedure as in Sect. 8.1 by constructing the cost functional

$$J(x_0, u(\cdot)) = \int_0^{\infty} \ell(x(s), u(s))ds,$$

where the running cost is of the same form as (8.2), which can be used to define the value function

$$V^*(x) = \inf_{u(\cdot)} J(x, u(\cdot)),$$

which satisfies the Hamilton-Jacobi-Bellman equation (HJB)

$$0 = \inf_{u \in \mathbb{R}^m} \left\{ L_f V^*(x) + L_F V^*(x)\theta + L_g V^*(x)u + \ell(x, u) \right\}.$$

Provided the value function is continuously differentiable (see Assumption 8.1), the optimal policy can then be derived from the HJB as

$$k^*(x) = -\tfrac{1}{2} R^{-1} L_g V^*(x)^\top. \tag{9.19}$$

Since the value function V^* is a CLF (see Theorem 8.1) and the optimal policy k^* is an explicit example of a stabilizing controller using V^* as a Lyapunov function, they can be used within the framework from the previous section to develop a hybrid control scheme that enforces satisfaction of an LTL specification. Of course, such an approach would require solving multiple HJB equations (one for each reach-avoid problem), which would quickly become computationally intractable for complex systems and specifications. Fortunately, the value function and policy for each optimal control problem can be safely learned online in real-time using the MBRL algorithm developed in Chap. 8.

Remark 9.2 In the special case that (4.1) is a linear system of the form $\dot{x} = Ax + Bu$, where the entries of A are possibly unknown and are treated as uncertain parameters, the value function of the optimal control problem for the resulting auxiliary system (9.18) is invariant to the choice of $x_d \in \mathbb{R}^n$. To see this, note that the feed-forward input is given by $u_d = -B^\top (BB^\top)^{-1} Ax_d$, implying that the auxiliary system (9.18) becomes $\dot{x} = Ax - BB^\top (BB^\top)^{-1} Ax_d + B\mu = Ax - Ax_d + B\mu = A\eta + B\mu$ for any choice of x_d. Provided the running cost is the same for any two optimal control problems characterized by (8.1), then the HJB is also the same for each problem, which implies they share the same value function. This will become important later on as, rather than learning a family of value functions, it is only necessary to learn one value function.

Recall that the MBRL approach proceeds by parameterizing the value function and optimal policy over a compact set $\mathcal{X} \subset \mathbb{R}^n$ as

$$\hat{V}(x, \hat{W}_c) = \hat{W}_c^\top \phi(x)$$
$$\hat{k}(x, \hat{W}_a) = -\tfrac{1}{2} R^{-1} g(x)^\top \tfrac{\partial \phi}{\partial x}(x)^\top \hat{W}_a,$$

where $\hat{W}_c, \hat{W}_a$ are the parameters and $\phi(x)$ is a vector of features, and then passing the approximated policy through a robust adaptive CBF (RaCBF) QP (8.20) to guarantee forward invariance of a safe set C. While the RaCBF-QP acts to shield the approximated policy from taking unsafe actions, an estimated model of the system dynamics, which is learned using the Concurrent Learning techniques from Sect. 4.2, is used to simulate potentially unsafe actions at various sample points to generate data for updating the value function and policy parameters. Conditions under which this MBRL converges to a neighborhood of the true value function and policy were introduced in Theorem 8.2 and stability of the origin under the learned policy is established in Theorem 8.4. A challenge with directly using the Lyapunov function from Theorem 8.4 in the CBF-stabilizable set framework introduced in the present chapter is that such a Lyapunov function guarantees convergence of a composite state trajectory consisting of the system states and weight estimation errors, which makes it challenging to develop precise bounds on the trajectory of the system itself. In the following result we provide an alternative method to establish stability of the closed-loop system under the learning-based policy $\hat{k}$ using only the value function as a Lyapunov function.

Theorem 9.2 *Consider system (4.1) under the influence of the learning-based policy (8.10). Suppose the estimated weights and parameters are updated according to (8.23), (8.24), (8.25), and (5.25). Let $\bar{\mathcal{B}}_r(0) \subset X$ be a closed ball of radius $r \in \mathbb{R}_{>0}$ contained within the compact set over which value function parameterization (8.8) is valid. Provided the conditions of Theorem 8.2 hold, and*

$$\iota := \gamma_3^{-1}(2\nu) < \gamma_2^{-1}(\gamma_1(r)),$$

where $\gamma_1, \gamma_2, \gamma_3 \in \mathcal{K}$ satisfy

$$\gamma_1(\|x\|) \le V^*(x) \le \gamma_2(\|x\|), \quad \gamma_3(\|x\|) \ge Q(x),$$

for all $x \in X$ and

$$\nu := \tfrac{1}{4}\overline{\|G_\varepsilon(x)\|_X} + \tfrac{1}{2}\overline{\left\| W^\top G_\phi(x) + \tfrac{\partial \varepsilon}{\partial x}(x) G_R(x)\tfrac{\partial \phi}{\partial x}(x)^\top \right\|}_X \bar{W}_a,$$

where $\bar{W}_a$ is a positive constant satisfying $\|\tilde{W}_a\| \le \bar{W}_a$ for all $t \ge 0$, then for any trajectory $t \mapsto x(t)$ with an initial condition such that

$$x(0) \le \gamma_2^{-1}(\gamma_1(r)),$$

there exists a time $T \in \mathbb{R}_{\ge 0}$ and a $\beta \in \mathcal{KL}$ such that

$$\|x(t)\| \le \beta(\|x_0\|, t), \qquad \forall t \in [0, T]$$
$$\|x(t)\| \le \gamma_1^{-1}(\gamma_2(\iota)), \qquad \forall t \in [T, \infty).$$

Proof Take V^* as a Lyapunov function candidate for the closed-loop system under the policy from (8.10). From (8.58), V^* can be upper bounded as

$$\dot{V}^* \leq -Q(x) + \frac{1}{4}\overline{\|G_\varepsilon(x)\|}_X + \frac{1}{2}\overline{\left\| W^\top G_\phi(x) + \frac{\partial \varepsilon}{\partial x}(x) G_R(x) \frac{\partial \phi}{\partial x}(x)^\top \right\|}_X \|\tilde{W}_a\|.$$

If the conditions of Theorem 8.2 hold, then $t \mapsto \tilde{W}_a(t)$ is bounded such that $\|\tilde{W}_a(t)\| \leq \bar{W}_a$ for all $t \in \mathbb{R}_{\geq 0}$. Hence, $\dot{V}^*$ can be further bounded as

$$\begin{aligned}
\dot{V} \leq & -Q(x) + \nu \\
\leq & -\gamma_3(\|x\|) + \nu \\
\leq & -\tfrac{1}{2}\gamma_3(\|x\|), \quad \forall \|x\| \geq \underbrace{\gamma_3^{-1}(2\nu)}_{\iota} .
\end{aligned} \tag{9.20}$$

Hence, provided that $\iota < \gamma_2^{-1}(\gamma_1(r))$ then Theorem 8.3 implies the existence of a $\beta \in \mathcal{KL}$ and a time $T \in \mathbb{R}_{\geq 0}$ such that for any initial condition $x_0 := x(0)$ satisfying $\|x_0\| \leq \gamma_2^{-1}(\gamma_1(r))$ the resulting solution $t \mapsto x(t)$ satisfies

$$\begin{aligned}
\|x(t)\| \leq & \beta(\|x_0\|, t), \qquad \forall t \in [0, T] \\
\|x(t)\| \leq & \gamma_1^{-1}(\gamma_2(\iota)), \qquad \forall t \in [T, \infty),
\end{aligned}$$

as desired.   □

The preceding theorem implies that all trajectories $t \mapsto x(t)$ starting in the set $\{x \in \mathcal{X} \mid \|x\| \leq \gamma_2^{-1}(\gamma_1(r))\}$ converge to the smaller set $\Omega := \{x \in \mathcal{X} \mid \|x\| \leq \gamma_1^{-1}(\gamma_2(\iota))\}$ in finite time and remain there for all times thereafter. Since the inverse of a class $\mathcal{K}$ function is a class $\mathcal{K}$ function and the composition of class $\mathcal{K}$ functions is again a class $\mathcal{K}$ function, the ultimate bound is a class $\mathcal{K}$ function of ι and therefore decreases with decreasing ι. Based on the proceeding Theorem, the ultimate bound can be decreased by decreasing $\overline{\|\nabla \varepsilon\|}_X$ (by choosing a more expressive function approximator), by decreasing $\lambda_{\min}(R)$ in the cost function, and by choosing a larger state penalty $Q(x)$ in the cost function. Hence, the ultimate bound can be made arbitrarily small through the appropriate selection of the cost function and function approximator. By taking the ultimate bound sufficiently small, the results of the previous sections can be directly extended to this setting, and then used within a hybrid system framework to develop controllers satisfying an LTL formula. More details on this extension are provided in the Notes at the end of this chapter.

9.5   Numerical Examples

Example 9.3 Our first example involves a persistent surveillance scenario for an uncertain system as in (4.1) with

$$\underbrace{\begin{bmatrix} \dot{x}_1 \\ \dot{x}_2 \end{bmatrix}}_{\dot{x}} = \underbrace{\begin{bmatrix} 0 \\ 0 \end{bmatrix}}_{f(x)} + \underbrace{\begin{bmatrix} x_1 & x_2 & 0 & 0 \\ 0 & 0 & x_1 & x_2 \end{bmatrix}}_{F(x)} \underbrace{\begin{bmatrix} \theta_1 \\ \theta_2 \\ \theta_3 \\ \theta_4 \end{bmatrix}}_{\theta} + \underbrace{\begin{bmatrix} 1 & 0 \\ 0 & 1 \end{bmatrix}}_{g(x)} \underbrace{\begin{bmatrix} u_1 \\ u_2 \end{bmatrix}}_{u}, \qquad (9.21)$$

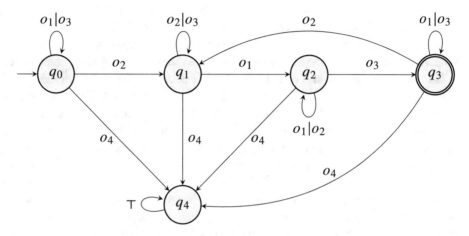

Fig. 9.2 DBA representing the specification in (9.22). The state q_0 is the initial state, q_3 is the accepting state, and q_4 is a "trap" state. If multiple transitions exist between states, only one transition, labeled by all observations that enable that transition is illustrated. For this simple example, the DTA for each state can be computed by inspection as $V_d(q_0) = 3$, $V_d(q_1) = 2$, $V_d(q_2)01$, $V_d(q_3) = 0$, $V_d(q_4) = \infty$

and an LTL specification

$$\varphi = \Box(\Diamond(o_2 \wedge \Diamond(o_1 \wedge \Diamond o_3))) \wedge \Box \neg o_4, \qquad (9.22)$$

where the regions corresponding to o_1, o_2, o_3 are areas of the state space that must be visiting infinitely often and the one corresponding to o_4 is a dangerous area that must avoided at all times. In words, (9.22) reads "always eventually visit $\mathcal{R}_{o_2}$ and then $\mathcal{R}_{o_1}$ and then $\mathcal{R}_{o_3}$ and always avoid $\mathcal{R}_{o_4}$." The DBA corresponding to (9.22) is displayed in Fig. 9.2. In accordance with minimizing the DTA over the transitions of the DBA in Fig. 9.2, an accepting word of ϕ can be computed as $w_o = (o_2 o_1 o_3)^\omega$. Thus, by Definition 9.3, to generate the accepting word of ϕ one needs to design a collection of feedback controllers that drive the continuous trajectory of the system through $\mathcal{R}_{o_2}$, then $\mathcal{R}_{o_1}$, then $\mathcal{R}_{o_3}$ infinitely often without ever entering $\mathcal{R}_{o_4}$. We consider each region as a circular disk that can be expressed as the zero sublevel set of $h_o(x) = \|x - x_o\|^2 - r_o^2$, where $r_o \in \mathbb{R}_{>0}$ denotes the radius of the region and $x_o \in \mathbb{R}^2$ its center. The true values of $\theta = [0.2 \ -0.3 \ 0.5 \ -0.5]^\top$ are assumed to be unknown to the controller, but the ranges of the parameters are assumed to be known in the sense that $\theta \in \Theta = [-1, 1]^4$. For simplicity, the uncertain parameters are identified using a modified version of the concurrent learning estimator (5.25), where, rather than integrating the dynamics to remove the dependence on $\dot{x}$, we simply assume that $\dot{x}$ is available for measurement. To synthesize a hybrid feedback controller capable of driving the system through the regions corresponding to the observations in w_o, we formulate a collection of optimal control problems with a quadratic cost defined by $Q(x) = \|x - x_d\|^2$ and $R = I_{2\times 2}$, where $x_d = x_o$ for each o. We parameterize the value function using a quadratic basis of the regulation error. Given that the system under consideration is linear, the corresponding auxiliary system from is also linear, implying that the HJB (8.4) simplifies to the algebraic Riccati equation, which can be solved using standard numerical tools to yield the true weights corresponding to the selected quadratic basis of the optimal control problem as $W = [1.208 \ -0.047 \ 0.624]^\top$. The parameters of the value function are updated using (8.23), (8.24) with $k_c = 1$, $\beta = 0.001$, whereas the parameters of the policy are updated using the simple projection-based update law from (8.26) with $k_a = 1$. To densely sample the system's operating region in accordance with the sufficient conditions of Lemma 8.1, the sample points used in the update laws are selected as the vertices of a 5×5 grid laid over $[-1.5, 1.5]^2 \subset \mathbb{R}^2$ yielding $N = 25$ extrapolation points. To guarantee that the system trajectory remains outside $\mathcal{R}_{o_4}$ we construct an RaCBF by taking $h(x) = h_{o_4}(x)$ and selecting the extended class $\mathcal{K}_\infty$ function as $\alpha(h(x)) = 10h(x)$. The initial value function and policy weights are selected as $\hat{W}_c(0) = \hat{W}_a(0) = [0.5 \ 0.5 \ 0.5]^\top$, the least squares matrix is initialized as $\Gamma(0) = 100I_{3\times3}$, and the initial drift parameters are set to zero. In the subsequent simulation we use a relatively low adaptation rate for updating the drift parameters to better illustrate the relationship between safety and adaptation.

The system is simulated for 100 s starting from an initial condition of $x = [1 \ 0]^\top$, the results of which are provided in Figs. 9.3 and 9.4. The plot in Fig. 9.3 illustrates the closed-loop system trajectory (denoted by the curve with varying color), where the system can be seen to visit the ROI in the corresponding order without ever entering $\mathcal{R}_{o_4}$ (red disk). Note that as time progresses the system is able to more closely approach the boundary of the safe set as the uncertain model parameters are identified. This phenomenon is further highlighted in Fig. 9.4b and c, which show the evolution of the estimated parameters and the value of the CBF over the duration of the simulation, respectively. At the start of the simulation the CBF-based controller accounts for the worst case model error, causing the safety margin to remain relatively high; however, as the estimation error decreases the value of the CBF is able to approach zero without crossing the boundary of the safe set as shown in Fig. 9.4e. Furthermore, the value function approximation scheme is able to closely approximate the solution to the optimal control problem as illustrated by the close convergence of the weights to their true values in Fig. 9.4d. By the end of the simulation, one of the true weights has been learned very closely, whereas the others can be seen to exhibit asymptotic convergence to a neighborhood of their true values.

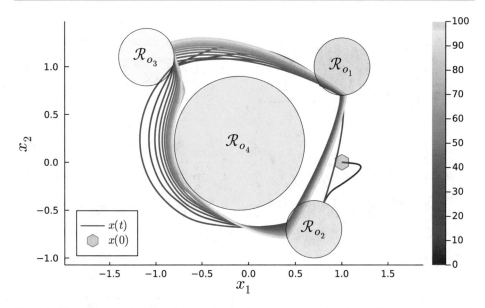

Fig. 9.3 Closed-loop trajectory of the uncertain linear system and regions of interest corresponding to observations in the specification. The curve of varying color denotes the system trajectory, where darker colors denote the system's state early in the simulation and lighter colors denote the system's state towards the end of the simulation. The purple, green, yellow, and red disks denote $\mathcal{R}_{o_1}, \mathcal{R}_{o_2}, \mathcal{R}_{o_3}, \mathcal{R}_{o_4}$, respectively. The initial condition of the system is represented as a blue hexagon

Example 9.4 Our next example involves a more complex nonlinear system in an obstacle-scattered environment subject to the specification

$$\varphi = \Box(\Diamond o_1 \wedge \Diamond o_2 \wedge \Diamond o_3 \wedge \Diamond o_4) \wedge \Box\neg(o_5 \vee o_6 \vee o_7 \vee o_8 \vee o_9),$$

which requires the system to visit the regions $\mathcal{R}_{o_1}, \mathcal{R}_{o_2}, \mathcal{R}_{o_3}, \mathcal{R}_{o_4}$ infinitely often and always avoid regions $\mathcal{R}_{o_5}, \mathcal{R}_{o_6}, \mathcal{R}_{o_7}, \mathcal{R}_{o_8}, \mathcal{R}_{o_9}$, where the regions are assumed to be of the same form as in the previous example. An accepting word for this specification can be computed as $w_o = (o_4 o_3 o_2 o_1)^\omega$. The system under consideration is a two-dimensional nonlinear control-affine system with parametric uncertainty and dynamics

$$\underbrace{\begin{bmatrix} \dot{x}_1 \\ \dot{x}_2 \end{bmatrix}}_{\dot{x}} = \underbrace{\begin{bmatrix} 0 \\ 0 \end{bmatrix}}_{f(x)} + \underbrace{\begin{bmatrix} x_1 \, x_2 \, 0 & 0 \\ 0 \ \ 0 \ \ x_1 & x_2(1 - (\cos(2x_1) + 2)^2) \end{bmatrix}}_{F(x)} \underbrace{\begin{bmatrix} -1 \\ 1 \\ -0.5 \\ -0.5 \end{bmatrix}}_{\theta} + \underbrace{\begin{bmatrix} 5 & 0 \\ 0 & 3 \end{bmatrix}}_{g(x)} \underbrace{\begin{bmatrix} u_1 \\ u_2 \end{bmatrix}}_{u},$$

where the set of possible parameter values Θ is chosen to be the same as in the previous example. The construction of the optimal control problem and all parameters associated

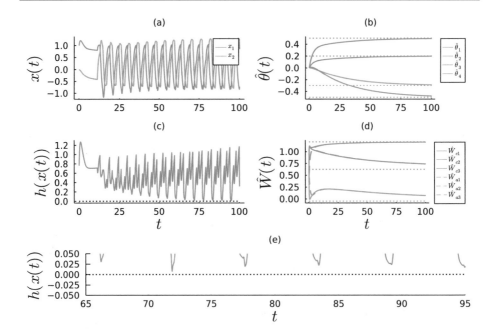

Fig. 9.4 Additional results from the simulation involving the uncertain linear system. The plot in Fig. 9.4a illustrates the evolution of the system states over time. Figure 9.4b displays the trajectory of the estimated drift weights (solid lines), where the dotted lines of corresponding color denote the true value of the weights. Figure 9.4c portrays the evolution of the CBF over time, where the blue curve denotes the value of the CBF along the system trajectory and the dotted black line represents $h(x) = 0$. A closer view of the CBF trajectory is provided in Fig. 9.4e, where the value is shown to be greater than zero for all time, indicating constraint satisfaction. The trajectory of the estimated value function and policy weights is provided in Fig. 9.4d, where the solid curves denote the estimated weights and the dotted lines denote the value of the ideal weights

with value function and policy approximation remain the same as in the previous example, where the initial weight estimates are chosen as $\hat{W}_c(0) = [1\ 1\ 1]^\top$, $\hat{W}_a(0) = 0.7\hat{W}_c(0)$ and the least squares matrix is initialized as $\Gamma(0) = 1000I_{3\times3}$ to ensure fast convergence to a stabilizing policy. To ensure the trajectory of the system satisfies the specification's safety requirements we take h_{o_i}, $i = 5, \ldots, 9$ as RaCBFs with extended class $\mathcal{K}_\infty$ functions $\alpha(h_{o_i}(x)) = 12h_{o_i}(x)$ for all $i = 5, \ldots, 9$, which are to construct an RaCBF safety filter. To demonstrate quick adaptation to uncertain dynamics, we use the same concurrent learning scheme as in the previous example, but increase the adaptation gain to achieve quicker parameter convergence.

The nonlinear system is simulated for 15 s from an initial condition of $x = [-2\ 0]^\top$, the results of which are provided in Figs. 9.5–9.6. Specifically, Fig. 9.5 illustrates the trajectory of the closed-loop system evolving an obstacle scattered environment, where the red disks denote the unsafe regions, which are avoided at all times. Similar to the previous example, varying the color of the curve in Fig. 9.5 is used to emphasize the passage of along the system's trajectory, which is shown to visit the regions of interest infinitely often. The periodic nature of the system's trajectory is further highlighted in Fig. 9.6a. The evolution of the estimated value function and policy weights are provided in Fig. 9.6d; however, a closed-form solution to the HJB for this problem is unavailable and thus accurately quantifying the quality of the approximation scheme is non-trivial. Despite this, the theoretical and numerical results clearly indicate that the MBRL algorithm is able to quickly adapt to each optimal control problem and safely navigate the system to a given target set in the presence of model uncertainty. The evolution of the estimated drift parameters is illustrated in Fig. 9.6b, where the estimated parameters are shown to rapidly converge to their true values. As a result, the RaCBF allows the system to closely approach the boundary of the safe set after about 1 s of simulation time. This behavior is further highlighted in Fig. 9.6c, which illustrates the minimum value among all RaCBFs at each timestep of the simulation. As shown in Fig. 9.6e, the minimum value among all RaCBFs remains non-negative for all time, indicating satisfaction of all safety requirements.

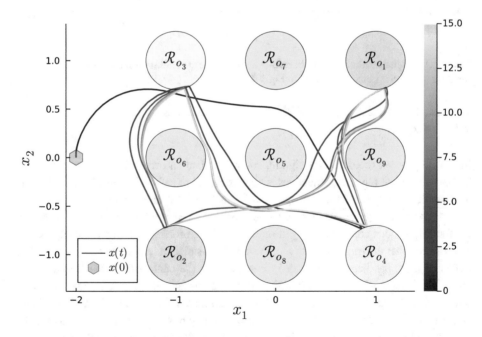

Fig. 9.5 Trajectory of the uncertain nonlinear system under the safe MBRL policy. Similar to Fig. 9.3, the curve of varying color denotes the trajectory of the system over time, and the disks in the state space represent various regions of interest, with red disks denoting unsafe areas

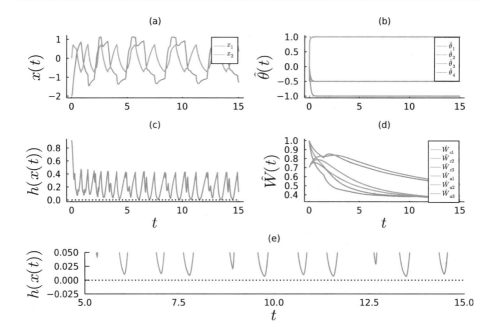

Fig. 9.6 Additional results from the simulation involving the uncertain nonlinear system. The subplots share the same interpretation as those in Fig. 9.4

9.6   Notes

In this chapter, we focused on designing controllers that enforce specifications richer than traditional control-theoretic objectives such as stability or safety. In particular, we focused on specifications given as linear temporal logic (LTL) formulas, which can be used to express complex requirements for dynamical systems. Temporal logics are used heavily in the formal methods community [1] in model checking problems in which the objective is to verify whether the trajectories of a discrete system satisfy a temporal logic specification. Recently, there has been an increasing interest from the control theoretic community in *formal synthesis* in which control policies for continuous dynamical systems are synthesized directly from a complex specification, such as an LTL formula. A central paradigm of early works in this field [2, 3] is the notion of *abstraction* in which a continuous-state dynamical system is viewed as a finite transition system whose dynamics capture only the essential behavior pertinent to the given specification. With such an abstraction in hand, a discrete plan can be obtaining by solving a Büchi or Rabin game over a product system composed of the abstraction and an automaton capturing the formal specification, which can then be executed by continuous control policies for the original system to ensure satisfaction of the formal specification. Tools that can be used to convert LTL formulas into a corresponding

Büchi automaton include `LTL2BA` [4] and `Spot` [5] Comprehensive textbooks that cover the abstraction-based approach to formal synthesis include [6, 7].

Although abstraction-based approaches provide strong formal guarantees of correctness, the computation of such abstractions is generally expensive and the class of systems for which continuous controllers can be synthesized is limited. To address these limitations, various researchers have proposed optimization-based techniques to formal synthesis (see [8] for a survey), in which the objective is optimizing a cost function subject to satisfaction of the temporal logic formula. An important component of such approaches is the use of temporal logics with *quantitative* semantics, such as signal temporal logic [9] that provides a metric of how well trajectories satisfy a given specification. In this setting, Boolean satisfaction of the specification can often be translated into a set of mixed-integer constraints and controller synthesis is performed by solving an optimal control problem with the objective of maximizing formula satisfaction subject to the aforementioned mixed integer constraints, resulting in a mixed integer convex programming (MICP) problem [10–12]. Such optimization-based approaches address some limitations of classical abstraction-based techniques, but can be computational intensive as large MICPs in real-time is challenging.

To cope with challenges of existing optimization-based approaches to control synthesis, various authors have attempted to leverage certificate-based functions, such as control barrier functions (CBFs), to enforce satisfaction of temporal logic formulas. In [13], time-varying CBFs are used to encode the satisfaction of an STL specification, allowing control inputs to be computed in a computationally efficient quadratic programming (QP) framework. Various extensions of the approach from [13] have been reported in [14–17]. Similar approaches to satisfying STL specifications using CBFs that ensure convergence to a set in finite time were developed in [18, 19]. Beyond STL, CBFs have also demonstrated success in developing controllers for other temporal logics, such at LTL [20, 21], and at bridging the gap between high-level planners outputting trajectories that satisfy an LTL specification and the low-level controllers that must be executed on the system to ensure satisfaction of the specification [22, 23].

Although the use of certificate-based functions has shown promise towards the development of computationally efficient control strategies that enforce temporal logic specifications, constructing such certificate functions for complex tasks may be challenging. An attractive alternative is to leverage learning-based techniques, such as reinforcement learning (RL), to obtain controllers capable of satisfying temporal logic specifications. Such approaches are often based upon using an LTL formula's corresponding automata to guide the learning process [24–29] or leverage quantitative semantics of certain temporal logics, such as STL, in the reward function of the RL problem to guide the learning process towards maximal satisfaction of the specification [30, 31].

The aforementioned approaches that use RL for temporal logic tasks typically do so in the offline episodic RL framework discussed in Chap. 8, which presents challenges in the context of safety-critical systems that must remain safe *during* the learning process. Initial attempts towards extending such ideas to online RL approaches were introduced in [32–34].

The method presented in the present chapter that extends the techniques from Chap. 8 and [35] to temporal logic tasks was introduced in [36]. The notion of a CBF-stabilizable set, which the development presented herein heavily relies upon, was first introduced in [37], where Theorem 9.1 was first stated and proved. The distance to acceptance function (DTA) was introduced in [38, 39] with similar approaches adopted in [40, 41].

References

1. Baier C, Katoen JP (2008) Principles of model checking. MIT Press
2. Tabuada P, Pappas G (2006) Linear time logic control of discrete-time linear systems. IEEE Trans Autom Control 51(12):1862–1877
3. Kloetzer M, Belta C (2008) A fully automated framework for control of linear systems from temporal logic specifications. IEEE Trans Autom Control 53(1):287–297
4. Gastin P, Oddoux D (2001) Fast ltl to büchi automata translation. In: Proceedings of the international conference on computer aided verification, pp 53–65
5. Duret-Lutz A, Lewkowicz A, Fauchille A, Michaud T, Renault E, Xu L (2016) Spot 2.0 – a framework for ltl and omega-automata manipulation. In: Proceedings of the international symposium on automated technology for verification and analysis, pp 122–129
6. Tabuada P (2009) Verification and control of hybrid systems: a symbolic approach. Spring Science & Business Media
7. Belta C, Yordanov B, Gol EA (2017) Formal methods for discrete-time dynamical systems. Springer
8. Belta C, Sadraddini S (2019) Formal methods for control synthesis: an optimization perspective. Ann Rev Control Robot Auton Syst 2:115–140
9. Maler O, Nickovic D (2004) Monitoring temporal properties of continuous signals. In: Lakhnech Y, Yovine S (eds) Formal techniques, modelling and analysis of timed and fault-tolerant systems. Springer, Berlin, Heidelberg, pp 152–166
10. Raman V, Donzé A, Maasoumy M, Murray RM, Sangiovanni-Vincentelli A, Seshia SA (2014) Model predictive control with signal temporal logic specifications. In: Proceedings of the IEEE conference on decision and control, pp 81–87
11. Raman V, Donzé A, Sadigh D, Murray RM, Seshia SA (2015) Reactive synthesis from signal temporal logic specifications. In: Proceedings of the international conference on hybrid systems: computation and control
12. Sadraddini S, Belta C (2015) Robust temporal logic model predictive control. In: Proceedings of the 53rd annual Allerton conference on communication, control, and computing, pp 772–779
13. Lindemann L, Dimarogonas DV (2019) Control barrier functions for signal temporal logic tasks. IEEE Control Syst Lett 3(1):96–101
14. Lindemann L, Dimarogonas DV (2019) Control barrier functions for multi-agent systems under conflicting local signal temporal logic tasks. IEEE Control Syst Lett 3(3):757–762
15. Lindemann L, Dimarogonas DV (2019) Decentralized control barrier functions for coupled multi-agent systems under signal temporal logic tasks. In: Proceedings of the European control conference, pp 89–94
16. Lindemann L, Dimarogonas DV (2020) Barrier function based collaborative control of multiple robots under signal temporal logic tasks. IEEE Trans Control Netw Syst 7(4):1916–1928
17. Gundana D, Kress-Gazit H (2021) Event-based signal temporal logic synthesis for single and multi-robot tasks. IEEE Robot Autom Lett 6(2):3687–3694

18. Garg K, Panagou D (2019) Control-lyapunov and control-barrier functions based quadratic program for spatio-temporal specifications. In: Proceedings of the IEEE conference on decision and control, pp 1422–1429

19. Xiao W, Belta C, Cassandras CG (2021) High order control lyapunov-barrier functions for temporal logic specifications. In: Proceedings of the American control conference, pp 4886–4891

20. Srinivasan M, Coogan S (2021) Control of mobile robots using barrier functions under temporal logic specifications. IEEE Trans Robot 37(2):363–374

21. Niu L, Clark A (2020) Control barrier functions for abstraction-free control synthesis under temporal logic constraints. In: Proceedings of the IEEE conference on decision and control, pp 816–823

22. Nilsson P, Ames AD (2018) Barrier functions: bridging the gap between planning from specifications and safety-critical control. In: Proceedings of the IEEE conference on decision and control, pp 765–772

23. Rosolia U, Singletary A, Ames AD (2022) Unified multirate control: from low-level actuation to high-level planning. IEEE Trans Autom Control 67(12):6627–6640

24. Sadigh D, Kim ES, Coogan S, Sastry SS, Seshia SA (2014) A learning based approach to control synthesis of markov decision processes for linear temporal logic specifications. In: Proceedings of the IEEE conference on decision and control, pp 1091–1096

25. Li X, Serlin Z, Yang G, Belta C (2019) A formal methods approach to interpretable reinforcement learning for robotic planning. Sci Robot 4(37)

26. Hasanbeig M, Kantaros Y, Abate A, Kroening D, Pappas GJ, Lee I (2019) Reinforcement learning for temporal logic control synthesis with probabilistic satisfaction guarantees. In: Proceedings of the IEEE conference on decision and control, pp 5338–5343

27. Bozkurt AK, Wang Y, Zavlanos MM, Pajic M (2020) Control synthesis from linear temporal logic specifications using model-free reinforcement learning. In: Proceedings of the IEEE international conference on robotics and automation, pp 10349–10355

28. Cai M, Hasanbeig M, Xiao S, Abate A, Kan Z (2021) Modular deep reinforcement learning for continuous motion planning with temporal logic. IEEE Robot Autom Lett 6(4):7973–7980

29. Cai M, Xiao S, Li B, Li Z, Kan Z (2021) Reinforcement learning based temporal logic control with maximum probabilistic satisfaction. In: Proceedings of the IEEE international conference on robotics and automation, pp 806–812

30. Aksaray D, Jones A, Kong Z, Schwager M, Belta C (2016) Q-learning for robust satisfaction of signal temporal logic specifications. In: Proceedings of the IEEE conference on decision and control, pp 6565–6570

31. Li X, Vasile CI, Belta C (2017) Reinforcement learning with temporal logic rewards. In: Proceedings of the IEEE/RSJ international conference on intelligent robots and systems, pp 3834–3839

32. Sun C, Vamvoudakis KG (2020) Continuous-time safe learning with temporal logic constraints in adversarial environments. In: Proceedings of the American control conference, pp 4786–4791

33. Kanellopoulos A, Fotiadis F, Sun C, Xu Z, Vamvoudakis KG, Topcu U, Dixon WE (2021) Temporal-logic-based intermittent, optimal, and safe continuous-time learning for trajectory tracking. arXiv:2104.02547

34. Cohen MH, Belta C (2021) Model-based reinforcement learning for approximate optimal control with temporal logic specifications. In: Proceedings of the international conference on hybrid systems: computation and control

35. Cohen MH, Belta C (2023) Safe exploration in model-based reinforcement learning using control barrier functions. Automatica 147:110684

36. Cohen MH, Serlin Z, Leahy KJ, Belta C (2023) Temporal logic guided safe model-based reinforcement learning: a hybrid systems approach. Nonlinear Anal: Hybrid Syst 47:101295

37. Cortez WS, Dimarogonas DV (2022) On compatibility and region of attraction for safe, stabilizing control laws. IEEE Trans Autom Control 67(9):4924–4931
38. Ding X, Belta C, Cassandras CG (2010) Receding horizon surveillance with temporal logic specifications. In: Proceedings of the IEEE conference on decision and control, pp 256–261
39. Ding X, Lazar M, Belta C (2014) Ltl receding horizon control for finite deterministic systems. Automatica 50:399–408
40. Bisoffi A, Dimarogonas DV (2018) A hybrid barrier certificate approach to satisfy linear temporal logic specifications. In: Proceedings of the American control conference, pp 634–639
41. Bisoffi A, Dimarogonas DV (2021) Satisfaction of linear temporal logic specifications through recurrence tools for hybrid systems. IEEE Trans Autom Control 66(2):818–825

Index

© The Editor(s) (if applicable) and The Author(s), under exclusive license to Springer
Nature Switzerland AG 2023
M. Cohen and C. Belta, *Adaptive and Learning-Based Control of Safety-Critical Systems*,
Synthesis Lectures on Computer Science,
https://doi.org/10.1007/978-3-031-29310-8

Printed in the United States
by Baker & Taylor Publisher Services